DIFFERENTIAL EQUATIONS and GROUP METHODS

for

SCIENTISTS and ENGINEERS

DIFFERENTIAL EQUATIONS and GROUP METHODS

for

SCIENTISTS and ENGINEERS

James M. Hill
Department of Mathematics,
The University of Wollongong,
Wollongong, N.S.W.
Australia

CRC Press
Boca Raton   Ann Arbor   London   Tokyo

Library of Congress Cataloging-in-Publication Data

Hill, James M.
 Differerential equations and group methods for scientists and engineers / James M. Hill.
   p. cm.
 Includes bibliographical references and index
 ISBN 0-8493-4442-5
 1. Differerential equations--Numerical solutions. 2. Transformation groups. I. Title.
 QA371.H57   1992
 515'.35--dc20   92-5860
                                                                                                CIP

  This book represents information obtained from authentic and highly regarded sources. Reprinted material is quoted with permission, and sources are indicated. A wide variety of references are listed. Every reasonable effort has been made to give reliable data and information, but the author and the publisher cannot assume responsibility for the validity of all materials or for the consequences of their use.

  All rights reserved. This book, or any parts thereof, may not be reproduced in any form without written consent from the publisher.

  Direct all inquiries to CRC Press, Inc., 2000 Corporate Blvd., N.W., Boca Raton, Florida, 33431.

© 1992 by CRC Press, Inc.

International Standard Book Number 0-8493-4442-5

Library of Congress Card Number 92-5860

Printed in the United States   1 2 3 4 5 6 7 8 9 0

Printed on acid-free paper

For Desley, Emily, Ruth and Thomas.

PREFACE

The trouble with solving differential equations is that whenever we are successful we seldom stop to ask why. The concept of one-parameter transformation groups which leave the differential equation invariant provides the only unified understanding of all known special solution techniques. In this book I have attempted to present a concise and self-contained account of the use of one-parameter groups to solve differential equations. The presentation is formal and is intended to appeal to Applied Mathematicians and Engineers whose principal concern is obtaining solutions of differential equations. I have included only the essentials of the subject, sufficient to enable the reader to attempt the group approach when solving differential equations. I have purposely not included all known results since this would inevitably lead to unnecessarily reproducing large portions of existing accounts. For example for ordinary differential equations, the account of the subject by L.E. Dickson, "Differential equations from the group standpoint", is still extremely readable and is recommended to the reader interested in pursuing the subject further. For partial differential equations the books by G.W. Bluman and J.D. Cole, "Similarity methods for differential equations", and by L.V. Ovsjannikov "Group properties of differential equations", contain several applications and examples which I have not reproduced here.

The first two chapters are introductory. Chapter 1 gives a general introduction with simple examples involving both ordinary and partial differential equations. In Chapter 2 the concepts of one-parameter groups and Lie series are introduced. Just as ordinary methods of solving differential equations often require a certain ingenuity so does the group approach. In order to establish some familiarity with the group method I have attempted to exploit our experience with linear equations. Most of us are aware that linear differential equations for $y(x)$ remain linear under the transformation $x_1 = f(x)$, $y_1 = g(x)y$ and Chapter 3 is devoted to implications of this result. In Chapters 4 and 5 I have tried to relate the usual theory for the group method with results obtained in the third chapter. In this respect the book differs from most accounts of the subject and I believe that a number of results given, especially in Chapter 3, are new.

The remaining two chapters are devoted to partial differential equations. For the most part the theory is illustrated with reference to diffusion related partial differential equations. The theory for linear partial differential equations is introduced in Chapter 6 for the classical diffusion or heat conduction equation and the Fokker-Planck equation. Non-linear equations are treated in Chapter 7. For partial differential equations the group approach is less satisfactory since for boundary value problems both the equation and boundary conditions must remain invariant. Here we principally consider only the invariance of the equation and view the group method as a means of systematically deducing solution types of a given partial differential equation.

This book forms the basis of a post-graduate course given at the University of Wollongong for a number of years. I have therefore included numerous examples and problems. In addition, I have used the problems at the end of each chapter to conveniently locate standard results for differential equations. On occasions I have also used these problems to include summaries of theory which is already adequately described in the literature. Several pages of answers and hints to the problems at the end of each chapter are included at the end of the book which are designed to produce a serviceable text for both undergraduate and graduate students in mathematics, science and engineering. In such disciplines, differential equations still play an important role and the present text serves to utilize the underlying hypothesis, which generate the particular differential equation, to actually solve the equation.

The existing theory of the solution of differential equations by means of one-parameter groups is by no means complete. Many of the inadequacies of the subject are highlighted in the text. When it does work it is very easy and it is therefore an area of knowledge which every Applied Mathematician ought to be aware of. Whatever the limitations of the group method may be, it will always represent a profoundly interesting idea towards solving differential equations. I hope this book proves to be useful and complements the existing literature.

The University of Wollongong, 1992.

TABLE OF CONTENTS

	Preface	vii
Chapter 1	Introduction	1
1.1	Introduction	1
1.2	Ordinary differential equations	1
1.3	Partial differential equations	5
1.4	Abel equations of the second kind	7
	Problems	13
Chapter 2	One-parameter groups and Lie series	17
2.1	Introduction	17
2.2	One-parameter transformation groups	17
2.3	Lie series and the Commutation theorem	22
	Problems	26
Chapter 3	Invariance of standard linear ordinary differential equations	31
3.1	Introduction	31
3.2	First order equation $y' + p(x)y = q(x)$	31
3.3	Second order homogeneous equation $y'' + p(x)y = 0$	34
3.4	Third order homogeneous equation $y''' + p(x)y' + q(x)y = 0$	37
3.5	Fourth order self-adjoint equation $y'''' + [p(x)y']' + q(x)y = 0$	40
	Problems	43

Table of contents

Chapter 4	First order ordinary differential equations	51
4.1	Introduction	51
4.2	Infinitesimal versions of y' and $y' = F(x,y)$ and the fundamental problem	52
4.3	Integrating factors and canonical coordinates for $y' = F(x,y)$	53
4.4	The alternative problem	57
4.5	The fundamental problem and singular solutions of $y' = F(x,y)$	60
4.6	Invariance of the associated first order partial differential equation	62
4.7	Lie's problem and area preserving groups	64
	Problems	68
Chapter 5	Second and higher order ordinary differential equations	79
5.1	Introduction	79
5.2	Infinitesimal versions of y'' and $y'' = F(x,y,y')$	79
5.3	Examples of the determination of $\xi(x,y)$ and $\eta(x,y)$	81
5.4	Determination of the most general differential equation invariant under a given one-parameter group	84
5.5	Applications	88
	Problems	93
Chapter 6	Linear partial differential equations	99
6.1	Introduction	99
6.2	Formulae for partial derivatives	100
6.3	Classical groups for the diffusion equation	102
6.4	Simple examples for the diffusion equation	103
6.5	Moving boundary problems	106
6.6	Fokker-Planck equation	109
6.7	Examples for the Fokker-Planck equation	114
6.8	Non-classical groups for the diffusion equation	118
	Problems	120

Table of contents   xi

Chapter 7	Non-linear partial differential equations	133
7.1	Introduction	133
7.2	Formulae for partial derivatives	135
7.3	Classical groups for non-linear diffusion	137
7.4	Non-classical groups for non-linear diffusion	142
7.5	Transformations of the non-linear diffusion equation	144
7.6	Similarity solutions of the non-linear diffusion equation	145
7.7	High order non-linear diffusion	151
	Problems	156
	Bibliography	167
	Answers and hints	169
	Summary of research areas	201

Chapter One
Introduction

1.1 Introduction

Although a good deal of research over the past two centuries has been devoted to differential equations our present understanding of them is far from complete. This book is concerned with obtaining solutions of differential equations by means of one-parameter transformation groups which leave the equation invariant. This subject was initiated by Sophus Lie over one hundred years ago. Such an approach is not always successful in deriving solutions. However it does provide a framework in which existing special methods of solution can be properly understood and also it is applicable to linear and non-linear equations alike. In formulating differential equations the Applied Mathematician inevitably makes certain assumptions. Using group theory these assumptions can be seen to hold the key to obtaining solutions of their equations.

The purpose of this chapter is to present a simple introduction to the subject for both ordinary and partial differential equations by means of simple familiar examples. For ordinary differential equations comprehensive accounts of the subject are given in the books by Cohen (1911), Page (1897) and more recently Bluman and Cole (1974) and in the review articles by Chester (1977) and Dickson (1924). For partial differential equations the reader may consult Bluman and Cole (1974) and Ovsjannikov (1967), both of which contain additional references. It is a curious fact that the process of utilizing one-parameter groups to solve differential equations tends to result in Abel equations of the second kind (namely (1.21)) for which there exists no general solution procedure. Interestingly enough the majority of Abel equations with integrals listed (see for example Murphy (1960)) frequently fall into one of three fairly trivial categories. Either they can be reduced to a standard linear first order equation with the substitution $z = y^2$ or they are of the standard first order homogeneous type or they can be readily reduced to a separable type. Accordingly we close this introductory chapter with one or two general details on Abel equations of the second kind.

1.2 Ordinary differential equations

In order to illustrate some of the ideas developed in this book we consider a simple example. It is well known that the 'homogeneous' first order differential equation

$$\frac{dy}{dx} = \frac{x^2 + y^2}{xy}, \tag{1.1}$$

can be made separable by the substitution $u(x,y) = y/x$, thus

$$u\,du = \frac{dx}{x},$$

which can be readily integrated to give

$$\log x - \frac{1}{2}\left(\frac{y}{x}\right)^2 = C, \tag{1.2}$$

where C denotes an arbitrary constant. We might well ask the following questions:

Question 1 Why does the substitution $u(x,y) = y/x$ lead to a separable equation for u?

Question 2 How do we interpret the degree of freedom embodied in the arbitrary constant C in the solution?

Answers to these questions can be provided within the framework of transformations which leave the differential equation unaltered. Consider the following transformation,

$$x_1 = e^{\epsilon} x, \quad y_1 = e^{\epsilon} y, \tag{1.3}$$

where ϵ is an arbitrary constant. We notice that (1.1) remains invariant under (1.3) in the sense that the differential equation in the new variables x_1 and y_1 is identical to the original equation, namely

$$\frac{dy_1}{dx_1} = \frac{x_1^2 + y_1^2}{x_1 y_1}. \tag{1.4}$$

Moreover we see that (1.3) satisfies the following:

(i) $\epsilon = 0$ gives the identity transformation $x_1 = x, \quad y_1 = y$,

(ii) $-\epsilon$ characterizes the inverse transformation $x = e^{-\epsilon} x_1, \quad y = e^{-\epsilon} y_1$,

(iii) if $x_2 = e^{\delta} x_1, \quad y_2 = e^{\delta} y_1$ then the product transformation is also a member of the set of transformations (1.3) and moreover is characterized by the parameter $\epsilon + \delta$, that is

$$x_2 = e^{\epsilon+\delta} x, \quad y_2 = e^{\epsilon+\delta} y.$$

A transformation satisfying these three properties is said to be a *one-parameter group of transformations*. We observe that the associativity law for groups follows from the property (iii). With this terminology established we might answer the above questions as follows:

Answer 1 The substitution $u(x,y) = y/x$ leads to a separable equation for u because $u(x,y)$ is an invariant of (1.3) in the sense that $u(x_1, y_1) = u(x,y)$ since,

$$u(x_1, y_1) = \frac{y_1}{x_1} = \frac{y}{x} = u(x,y), \tag{1.5}$$

Introduction 3

and it is this property which results in a simplification of (1.1). In general we shall see that if a differential equation is invariant under a one-parameter group of transformations then use of an invariant of the group results in a simplification of the differential equation. If the differential equation is of first order then it becomes separable while if the equation is of higher order then use of an invariant of the group permits a reduction in the order of the equation by one.

Answer 2 From (1.2) and (1.3) we see that we have

$$\log x_1 - \frac{1}{2}\left(\frac{y_1}{x_1}\right)^2 = C + \epsilon, \tag{1.6}$$

so that the degree of freedom in the solution (1.2) resulting from the arbitrary constant C is related to the invariance of the differential equation (1.1) under the group of transformations (1.3) which is characterized by the arbitrary parameter ϵ. That is, the transformation (1.3) permutes the solution curves (1.2). In general we shall see that for every one-parameter group in two variables there are functions $u(x, y)$ and $v(x, y)$ such that the group becomes

$$u(x_1, y_1) = u(x, y), \quad v(x_1, y_1) = v(x, y) + \epsilon. \tag{1.7}$$

Moreover if a first order differential equation is invariant under this group then in terms of these new variables u and v it takes the form,

$$\frac{dv}{du} = \phi(u), \tag{1.8}$$

and consequently has a solution of the form

$$v + \psi(u) = C, \tag{1.9}$$

for appropriate functions $\phi(u)$ and $\psi(u)$.

Example 1.1 Integrate the differential equation

$$y\frac{dy}{dx} = \frac{2}{x^3} - \frac{3}{x^2}y.$$

This is an Abel equation of the second kind (Murphy (1960), page 25) which is not readily amendable to any of the standard devices. However the equation is clearly invariant under the group

$$x_1 = e^\epsilon x, \quad y_1 = e^{-\epsilon} y,$$

and therefore we choose $u(x, y) = xy$ as the new dependent variable and the differential equation becomes,

$$xu\frac{du}{dx} = u^2 - 3u + 2,$$

which is separable and can therefore be readily integrated as follows,

$$\frac{dx}{x} = \frac{u\,du}{(u-1)(u-2)} = \left\{\frac{2}{(u-2)} - \frac{1}{(u-1)}\right\} du,$$

so that

$$(xy-2)^2 = Cx(xy-1),$$

gives the required integral.

It is worthwhile emphasizing that not all equations can be solved in such a simple manner. Consider for example,

$$y\frac{dy}{dx} = \left(\frac{2}{x^3} + 6\right) - \left(\frac{3}{x^2} + 6x\right) y,$$

which arises in finite elasticity. This equation is again an Abel equation of the second kind but in this case there is apparently no simple group which leaves the equation invariant. In this general introduction it may be appropriate to mention two related areas for which group theory has not yet been applied. The reader might well like to bear these problems in mind with a view to developing results in these areas.

(a) Differential-difference equations

It is well known that formal solutions of linear differential-difference equations, for example

$$\frac{dy(x)}{dx} = -y(x - x_0), \tag{1.10}$$

where x_0 is a constant, can be expressed as

$$y(x) = \sum_j C_j e^{-\omega_j x},$$

where C_j are arbitrary constants and ω_j denote the roots of $\omega = e^{\omega x_0}$. If the equation is non-linear then there are no such general methods of solution. Consider for example Hutchinson's equation which can be written as

$$\frac{dy(x)}{dx} = y(x)[1 - y(x - x_0)], \tag{1.11}$$

and arises in theory of populations. What are the implications of group theory, if any, for equations of this type? (See Problems 19 and 20 of Chapter 4).

Introduction 5

(b) **Differential equations invariant under transformations which cannot be characterized as one-parameter groups**

A differential equation occurring in fluid dynamics is

$$\frac{d^2x}{dt^2} = 2\frac{dx}{dt} + \frac{(5+3x)}{4x(1+x)}\left(\frac{dx}{dt}\right)^2 + \frac{3x(1-x)}{(1+x)}. \tag{1.12}$$

It can be verified that if $x(t)$ is a solution then so is $x(t)^{-1}$ since with $X(t) = x(t)^{-1}$ and using

$$\frac{dX}{dt} = -\frac{1}{x^2}\frac{dx}{dt}, \quad \frac{d^2X}{dt^2} = -\frac{1}{x^2}\frac{d^2x}{dt^2} + \frac{2}{x^3}\left(\frac{dx}{dt}\right)^2,$$

we have

$$\frac{d^2X}{dt^2} - 2\frac{dX}{dt} - \frac{(5+3X)}{4X(1+X)}\left(\frac{dX}{dt}\right)^2 - \frac{3X(1-X)}{(1+X)}$$

$$= -\frac{1}{x^2}\left\{\frac{d^2x}{dt^2} - 2\frac{dx}{dt} - \frac{(5+3x)}{4x(1+x)}\left(\frac{dx}{dt}\right)^2 - \frac{3x(1-x)}{(1+x)}\right\}$$

$$= 0.$$

If we now let $y = dx/dt$ then (1.12) becomes

$$y\frac{dy}{dx} = \frac{3x(1-x)}{(1+x)} + 2y + \frac{(5+3x)}{4x(1+x)}y^2, \tag{1.13}$$

which is again an Abel equation of the second kind. From the solution property of (1.12) we can deduce that (1.13) remains invariant under the transformation

$$x_1 = \frac{1}{x}, \quad y_1 = -\frac{y}{x^2}, \tag{1.14}$$

which cannot be characterized as a one-parameter group. Can we use such invariance properties to determine solutions of differential equations?

1.3 Partial differential equations

Unlike ordinary differential equations the success of the group approach for partial differential equations depends to a considerable extent on the accompanying boundary conditions. That is, the group approach is only effective in the solution of boundary value problems if both the equation and boundary conditions are left unchanged by the one-parameter group. For the most part we confine our attention to specific differential equations rather than boundary value problems. For any particular boundary value problem we should always first look for any simple invariance properties. These may be

6 *Differential Equations and Group Methods for Scientists and Engineers*

more apparent from the physical hypothesis of the problem rather than its mathematical formulation. If no such invariance can be found and if the problem merits a numerical solution then the group approach might still be relevant as a means of checking the numerical technique with artificially imposed boundary conditions which permit an exact analytic solution.

As an illustration we consider a boundary value problem for which both the partial differential equation and the boundary conditions are invariant under a simple one-parameter group.

Example 1.2 Determine the source solution for the one-dimensional diffusion or heat conduction equation for $c(x,t)$, namely

$$\frac{\partial c}{\partial t} = \frac{\partial^2 c}{\partial x^2} \quad (t > 0, -\infty < x < \infty). \tag{1.15}$$

The source solution for (1.15) is the solution which vanishes at infinity for all times and initially satisfies

$$c(x,0) = c_0 \delta(x), \tag{1.16}$$

where c_0 is a constant specifying the strength of the source and $\delta(x)$ is the usual Dirac delta function. We observe that both of (1.15) and (1.16) are left unchanged by the transformation

$$x_1 = e^{\epsilon} x, \quad t_1 = e^{2\epsilon} t, \quad c_1 = e^{-\epsilon} c, \tag{1.17}$$

where ϵ denotes an arbitrary constant and we have made use of the elementary property of delta functions,

$$\delta(\lambda x) = \lambda^{-1} \delta(x),$$

for any non-zero constant λ. Thus if $c = \phi(x,t)$ is the solution of (1.15) and (1.16) then we have also $c_1 = \phi(x_1, t_1)$. Clearly this is the case if $\phi(x,t)$ has the functional form

$$\phi(x,t) = t^{-1/2} \psi(xt^{-1/2}), \tag{1.18}$$

for some function ψ of the argument indicated. Upon substituting (1.18) into (1.15) we obtain the ordinary differential equation

$$2\psi''(\xi) + \xi \psi'(\xi) + \psi(\xi) = 0, \tag{1.19}$$

where ξ denotes $xt^{-1/2}$ and primes indicate differentiation with respect to ξ. Equation (1.19) can be integrated immediately to give

$$2\psi'(\xi) + \xi \psi(\xi) = 2A,$$

Introduction 7

where A is a constant and a further integration of this first order equation yields

$$\psi(\xi) = Ae^{-\xi^2/4} \int^{\xi} e^{\tau^2/4} d\tau + Be^{-\xi^2/4},$$

where B denotes a further constant. Thus the appropriate solution of (1.19) vanishing at infinity is simply

$$\psi(\xi) = Be^{-\xi^2/4}, \tag{1.20}$$

where the constant is determined from (1.16), namely

$$\int_{-\infty}^{\infty} c(x,t)dx = c_0.$$

From this equation, (1.18) and (1.20) we find that the required solution of the boundary value problem (1.15) and (1.16) becomes

$$c(x,t) = c_0 \frac{e^{-x^2/4t}}{(4\pi t)^{1/2}} \quad (t > 0, \quad -\infty < x < \infty),$$

which is of course well known.

For our purposes this example serves firstly as a specific non-trivial boundary value problem for which both the differential equation and boundary conditions are invariant under a one-parameter group. Secondly it serves to illustrate that knowledge of a one-parameter group leaving the equation invariant enables, at least in the case of two independent variables, the partial differential equation to be reduced to an ordinary differential equation. For more independent variables knowledge of a group leaving the equation unchanged reduces the number of independent variables by one. In this book we give the general procedure for determining the group such as (1.17) which leaves a specific equation invariant. We also give the general technique for establishing the functional form of the solution such as that given by (1.18).

1.4 Abel equations of the second kind

We shall see repeatedly that a consequence of attempting to solve differential equations by means of one-parameter groups is the appearance of Abel equations of the second kind, namely first order ordinary differential equations of the form,

$$y\frac{dy}{dx} + a(x) + b(x)y = 0, \tag{1.21}$$

where $a(x)$ and $b(x)$ are given functions of x. In general this equation is not amenable to standard devices and a general integral for arbitrary functions $a(x)$ and $b(x)$ is not known. Such a result would have many implications in Applied Mathematics and accordingly the interested reader is alerted that here is a problem worthy of their attention. Because this is an equation which will occur again and again we present in this section a synopsis of the simple results relating to its origins and known solutions.

Equation (1.21) can be seen to arise immediately from non-linear oscillation theory where the function $x = x(t)$ satisfies a second order ordinary differential equation of the form
$$\frac{d^2x}{dt^2} + b(x)\frac{dx}{dt} + a(x) = 0.$$
On making use of the standard substitution,
$$y = \frac{dx}{dt}, \quad y\frac{dy}{dx} = \frac{d^2x}{dt^2},$$
we may readily deduce the Abel equation (1.21). Actually Abel originally introduced (1.21) by means of the equation
$$[y + s(x)]\frac{dy}{dx} + [p(x) + q(x)y + r(x)y^2] = 0, \tag{1.22}$$
where $p(x), q(x), r(x)$ and $s(x)$ are known functions of x. Abel showed that the change of variable,
$$z = [y + s(x)]e^{t(x)}, \quad t(x) = \int^x r(\tau)d\tau,$$
reduces (1.22) to an equation of the form (1.21), namely
$$z\frac{dz}{dx} + \left\{(p - qs + rs^2)e^{2t} + \left(q - 2rs - \frac{ds}{dx}\right)ze^t\right\} = 0, \tag{1.23}$$
for which there are clearly two special cases to consider:

(a) $p - qs + rs^2 = 0$, in this case $(y+s)$ is a factor of (1.22) and therefore (1.22) simplifies to the standard linear first order ordinary differential equation (namely equation (3.3)),

(b) $q = 2rs + ds/dx$, in this case (1.22) can actually be written as the standard linear first order ordinary differential equation with dependent variable $(y + s)^2$, that is
$$\frac{d}{dx}(y+s)^2 + 2r(y+s)^2 = 2\left\{s\left(\frac{ds}{dx} + rs\right) - p\right\}.$$

Another first order differential equation giving rise to (1.21) is known as the Abel equation of the first kind,
$$\frac{dy}{dx} = P(x) + Q(x)y + R(x)y^2 + S(x)y^3, \tag{1.24}$$
where $P(x), Q(x), R(x)$ and $S(x)$ are all known functions of x. If $y_1(x)$ is a known special solution of (1.24) then the substitution
$$w(x) = \frac{e^{T(x)}}{[y(x) - y_1(x)]}, \quad T(x) = \int^x [Q(\tau) + 2R(\tau)y_1(\tau) + 3S(\tau)y_1(\tau)^2]d\tau,$$
reduces (1.24) to an equation of the form (1.21), that is,
$$w\frac{dw}{dx} + S(x)e^{2T(x)} + [3S(x)y_1(x) + R(x)]we^{T(x)} = 0.$$

Introduction

This is most easily seen by substituting
$$y(x) = y_1(x) + \frac{e^{T(x)}}{w(x)},$$
into (1.24) and using the fact that $y_1(x)$ is a known special solution.

Example 1.3 With the notation
$$A(x) = \int^x a(\tau)d\tau, \quad B(x) = \int^x b(\tau)d\tau,$$
and writing (1.21) in the form
$$ydy + [a(x) + b(x)y]dx = 0, \tag{1.25}$$
verify the following integrating factors $\mu(x,y)$ for the various special cases listed:

(i) $\quad \mu(x,y) = \frac{\exp\{[y+B(x)]^2\}}{B(x)}, \qquad a(x) = -\frac{b(x)}{2B(x)},$

(ii) $\quad \mu(x,y) = \left[y + \frac{(\alpha+1)}{\alpha}B(x)\right]^\alpha, \qquad a(x) = -\frac{(\alpha+1)}{\alpha^2}b(x)B(x),$

(iii) $\quad \mu(x,y) = \exp\left\{\frac{2B(x)^2}{[B(x)^2+2yB(x)+\alpha]}\right\}, \qquad a(x) = \frac{b(x)[B(x)^2+\alpha]^2}{4B(x)^3},$

where α denotes an arbitrary constant.

If $\mu(x,y)$ is an integrating factor for (1.25) then we require that
$$\frac{\partial}{\partial x}(\mu y) = \frac{\partial}{\partial y}\{\mu[a(x) + b(x)y]\},$$
so that
$$y\frac{\partial \mu}{\partial x} - [a(x) + b(x)y]\frac{\partial \mu}{\partial y} = b(x)\mu.$$

Thus we need to verify that this equation is satisfied for each of the special cases listed.

(i) $\quad \mu(x,y) = \frac{\exp\{[y+B(x)]^2\}}{B(x)}, \qquad a(x) = \frac{-b(x)}{2B(x)}.$

In this case we have

$$y\frac{\partial \mu}{\partial x} - [a(x) + b(x)y]\frac{\partial \mu}{\partial y} - b(x)\mu$$

$$= e^{(y+B)^2}\left\{y\left[2(y+B)\frac{b}{B} - \frac{b}{B^2}\right] - (a+by)\frac{2}{B}(y+B) - \frac{b}{B}\right\}$$

$$= e^{(y+B)^2}\frac{(y+B)}{B}\left\{2yb - 2(a+by) - \frac{b}{B}\right\}$$

$$= -e^{(y+B)^2}\frac{(y+B)}{B^2}(b + 2aB),$$

which is zero if $a(x)$ and $b(x)$ are such that $b + 2aB = 0$.

(ii) $$\mu(x,y) = \left[y + \frac{(\alpha+1)}{\alpha}B(x)\right]^\alpha, \quad a(x) = -\frac{(\alpha+1)}{\alpha^2}b(x)B(x).$$

In this case we have

$$y\frac{\partial \mu}{\partial x} - [a(x) + b(x)y]\frac{\partial \mu}{\partial y} - b(x)\mu$$

$$= \left[y + \frac{(\alpha+1)}{\alpha}B\right]^{\alpha-1}\left\{(\alpha+1)by - (a+by)\alpha - b\left[y + \frac{(\alpha+1)}{\alpha}B\right]\right\}$$

$$= -\left[y + \frac{(\alpha+1)}{\alpha}B\right]^{\alpha-1}\left\{\alpha a + \frac{(\alpha+1)}{\alpha}bB\right\},$$

which is again zero with the stated condition.

(iii) $$\mu(x,y) = \exp\left\{\frac{2B(x)^2}{[B(x)^2 + 2yB(x) + \alpha]}\right\}, \quad a(x) = \frac{b(x)[B(x)^2 + \alpha]^2}{4B(x)^3}.$$

In this case we have

$$y\frac{\partial \mu}{\partial x} - [a(x) + b(x)y]\frac{\partial \mu}{\partial y} - b(x)\mu$$

$$= \exp\left\{\frac{2B^2}{(B^2+2yB+\alpha)}\right\}\left\{\frac{4bBy}{(B^2+2yB+\alpha)} - \frac{4bB^2(B+y)y}{(B^2+2yB+\alpha)^2} + \frac{(a+by)4B^3}{(B^2+2yB+\alpha)^2} - b\right\}$$

$$= \exp\left\{\frac{2B^2}{(B^2+2yB+\alpha)}\right\}\left\{\frac{4B^3a - b(B^2+\alpha)^2}{(B^2+2yB+\alpha)^2}\right\},$$

and again this is zero with the stated constraint on the functions $a(x)$ and $b(x)$.

These integrating factors are due to Abel and further results, also due to him, are described in Problem 15.

Example 1.4 Show that equation (1.22) can be made separable by means of the transformation
$$y = s(x)u,$$
provided that the functions $p(x)$ and $q(x)$ are such that
$$p(x) = \alpha[s'(x) + r(x)s(x)]s(x), \quad q(x) = \beta s'(x) + (\beta + 1)r(x)s(x),$$
where α and β denote arbitrary constants and primes denote differentiation with respect to x. Further with $p(x)$ and $q(x)$ so defined derive the corresponding Abel equation of the second kind (namely, (1.21)) and show that
$$a(x) = -\frac{(\beta - \alpha)}{(\beta - 1)^2} b(x) B(x),$$
which is essentially case (ii) of Example 1.3.

From $y = s(x)u$ and $p(x)$ and $q(x)$ as given above we see that (1.22) becomes
$$(1 + u)s\left(s\frac{du}{dx} + s'u\right) + \alpha(s' + rs)s + [\beta s' + (\beta + 1)rs]su + rs^2u^2 = 0,$$
which simplifies to give
$$(1 + u)\frac{du}{dx} + \left(r + \frac{s'}{s}\right)[u^2 + (\beta + 1)u + \alpha] = 0,$$
which is clearly separable. For the second part we have from equation (1.23)
$$a(x) = (p - qs + rs^2)e^{2t}, \quad b(x) = (q - 2rs - s')e^t,$$
and the appropriate restrictions on $p(x)$ and $q(x)$ give
$$a(x) = (\alpha - \beta)[s'(x) + r(x)s(x)]s(x)e^{2t(x)},$$
$$b(x) = (\beta - 1)[s'(x) + r(x)s(x)]e^{t(x)},$$
and therefore by division we have
$$\frac{a(x)}{b(x)} = -\frac{(\beta - \alpha)}{(\beta - 1)} s(x)e^{t(x)}.$$
But using the above expression for $b(x)$ we have
$$B(x) = \int^x b(\tau)d\tau = (\beta - 1)s(x)e^{t(x)},$$
from which the given constraint may be deduced.

Example 1.5 For equation (1.22), with $p(x)$ and $q(x)$ as defined in the previous example, show that the differential equation (1.22) remains invariant under the transformation of the form,

$$x_1 = f(x), \quad y_1 = g(x)y,$$

which is given by

$$s(x_1)e^{t(x_1)} = s(x)e^{t(x)+\epsilon}, \quad \frac{y_1}{s(x_1)} = \frac{y}{s(x)}, \qquad (1.26)$$

where $t(x)$ is as previously defined.

This invariance property of (1.22) is most easily seen by noting that

$$u = \frac{y}{s(x)} = \frac{y_1}{s(x_1)},$$

is an invariant of the transformation and that from the previous example we have that (1.22) simplifies to give

$$\frac{(1+u)du}{[u^2 + (\beta+1)u + \alpha]} = -\left[r(x) + \frac{s'(x)}{s(x)}\right]dx. \qquad (1.27)$$

It follows that (1.22) is unchanged by the transformation (1.26), that is

$$[y_1 + s(x_1)]\frac{dy_1}{dx_1} + [p(x_1) + q(x_1)y_1 + r(x_1)y_1^2] = 0,$$

provided that we are able to show that the right-hand side of (1.27) is invariant, namely

$$\left[r(x) + \frac{s'(x)}{s(x)}\right]dx = \left[r(x_1) + \frac{s'(x_1)}{s(x_1)}\right]dx_1,$$

which can be readily verified by taking logarithms of (1.26)$_1$ and differentiating. Hence for $p(x)$ and $q(x)$ as given in the previous example we have established (1.22) remains unchanged by the transformation (1.26).

PROBLEMS

1. Determine in each case the constants α and β such that the one-parameter group

$$x_1 = e^{\alpha \epsilon} x, \quad y_1 = e^{\beta \epsilon} y,$$

leaves the following differential equations invariant. Use an invariant of the group to integrate the equation.

   (i) $\frac{dy}{dx} = \frac{A}{x^{3/2}} + By^3$ (A and B are constants),

   (ii) $x\left(x^4 - 2y^3\right)\frac{dy}{dx} + \left(2x^4 + y^3\right)y = 0$,

   (iii) $x\left(A + xy^n\right)\frac{dy}{dx} + By = 0$ (A, B and n are constants).

2. Verify that,

$$x_1 = x + \epsilon, \quad y_1 = e^{-2\epsilon} y,$$

is a one-parameter group of transformations and hence integrate the differential equation

$$(1 - 2x - \log y)\frac{dy}{dx} + 2y = 0.$$

3. Integrate the differential equation

$$(x - y)^2 \frac{dy}{dx} = A^2 \quad (A \text{ is a constant}),$$

by observing that the equation admits the group

$$x_1 = x + \epsilon, \quad y_1 = y + \epsilon.$$

4. Given that $\rho(x)$ is a solution of the linear differential-difference equation (1.10) show that

$$y(x) = \frac{\rho(x - x_0)}{\rho(x)},$$

is a solution of the non-linear differential-difference equation

$$\frac{dy(x)}{dx} = y(x)[y(x) - y(x - x_0)].$$

5. Show that the transformation

$$y(x) = \frac{e^x}{f(e^{x-x_0})},$$

reduces equation (1.11) to the differential equation

$$\frac{df(t)}{dt} = \frac{f(t)}{f(\lambda t)},$$

where $t = e^{x-x_0}$ and $\lambda = e^{-x_0}$.

6. Show that with $\omega = y/x$ the differential equation (1.13) becomes

$$x\omega \frac{d\omega}{dx} = \frac{3(1-x)}{(1+x)} + 2\omega + \frac{(1-x)}{(1+x)}\frac{\omega^2}{4}.$$

Show further that the substitution $s = (1-x)/(1+x)$ yields

$$\frac{(s^2-1)}{2}\omega\frac{d\omega}{ds} = 3s + 2\omega + \frac{s\omega^2}{4},$$

and observe that the transformation (1.14) becomes $\omega_1 = -\omega$ and $s_1 = -s$.

7. Observe that the partial differential equation (1.15) remains invariant under the transformation

$$x_1 = e^\epsilon x, \quad t_1 = e^{2\epsilon}t, \quad c_1 = c,$$

so that the equation admits solutions of the form $c(x,t) = \phi(xt^{-1/2})$. Deduce the ordinary differential equation for ϕ and hence show that

$$c(x,t) = A\int_0^{xt^{-1/2}} e^{-y^2/4}dy + B,$$

where A and B denote arbitrary constants.

8. **Continuation.** For the non-linear diffusion equation

$$\frac{\partial c}{\partial t} = \frac{\partial}{\partial x}\left(D(c)\frac{\partial c}{\partial x}\right),$$

where the diffusivity D is a function of c only, use the one-parameter group and functional form of the solution in the previous problem to deduce the ordinary differential equation

$$D(\phi)\phi''(\xi) + \frac{dD(\phi)}{d\phi}\phi'(\xi)^2 + \frac{\xi\phi'(\xi)}{2} = 0.$$

9. **Continuation.** For the case $D(c) = c$ show that the ordinary differential equation of the previous problem remains invariant under the group

$$\xi_1 = e^\epsilon \xi, \quad \phi_1 = e^{2\epsilon}\phi,$$

and that with $\psi = \phi/\xi^2$ the equation reduces to the Abel equation of the second kind

$$\psi p\frac{dp}{d\psi} + p^2 + \left(7\psi + \frac{1}{2}\right)p + \psi(6\psi + 1) = 0,$$

where $p = d\psi/dy$ and $y = \log \xi$. Show that the singular solution $p = -2\psi$ corresponds to the solution $\phi(\xi)$ constant.

10. **Continuation.** Show that the transformation $p = q/\psi$ reduces the Abel equation in the previous problem to

$$q\frac{dq}{d\psi} + \left(7\psi + \frac{1}{2}\right)q + \psi^2(6\psi + 1) = 0,$$

which is in standard form for an Abel equation of the second kind.

11. By making the substitution $z = e^{Bc}$, show that the non-linear diffusion equation

$$\frac{\partial c}{\partial t} = \frac{\partial}{\partial x}\left(Ae^{Bc}\frac{\partial c}{\partial x}\right),$$

where A and B denote constants, yields

$$\frac{\partial z}{\partial t} = Az\frac{\partial^2 z}{\partial x^2}.$$

Show that this equation remains invariant under the group

$$z_1 = e^{m\epsilon}z, \quad t_1 = e^{n\epsilon}t, \quad x_1 = e^{\epsilon}x,$$

provided that the constants m and n are such that

$$m + n = 2.$$

12. **Continuation.** Show that the equation for z given in the previous problem admits a solution of the form

$$z(x,t) = t^{m/n}\phi(\xi),$$

where $\xi = x/t^{1/n}$ and where $\phi(\xi)$ satisfies the ordinary differential equation

$$nA\phi(\xi)\phi''(\xi) + \xi\phi'(\xi) - m\phi(\xi) = 0.$$

Show further that this equation also remains unaltered by the one-parameter group given in Problem 9 and that the Abel equation of the second kind may be deduced,

$$p\frac{dp}{d\psi} + \left(2\psi + \frac{1}{A}\right) + \left(3 + \frac{1}{nA\psi}\right)p = 0,$$

where p and ψ are exactly as defined in Problem 9.

13. By looking for solutions of the non-linear diffusion equation

$$\frac{\partial c}{\partial t} = \frac{\partial}{\partial x}\left(Ae^{Bc}\frac{\partial c}{\partial x}\right),$$

where A and B are constants, of the form

$$c(x,t) = f(x) + g(t),$$

deduce the solution

$$c(x,t) = \frac{1}{B} \log \left\{ \frac{(x-x_0)^2 + C}{2A(t_0 - t)} \right\},$$

where x_0, t_0 and C denote arbitrary constants.

14. By calculating the quantity,

$$\frac{\partial c_1}{\partial t_1} - \frac{\partial^2 c_1}{\partial x_1^2},$$

show directly, using the chain rule for partial derivatives, that the classical diffusion equation (1.15) remains invariant under the following transformations,

(i) $\quad x_1 = x + \epsilon t, \quad t_1 = t, \quad c_1 = c \exp\left(-\frac{\epsilon x}{2} - \frac{\epsilon^2 t}{4}\right),$

(ii) $\quad x_1 = \dfrac{x}{(1-\epsilon t)}, \quad t_1 = \dfrac{t}{(1-\epsilon t)}, \quad c_1 = c(1-\epsilon t)^{1/2} \exp\left(-\dfrac{\epsilon x^2}{4(1-\epsilon t)}\right).$

15. In the notation of Example 1.3 verify the following integrating factors for the Abel equation of the second kind (1.25), subject to the stated restriction on $a(x)$ and $b(x)$, thus

(i) $\quad \mu(x,y) = \exp\left\{ [y + B(x)]^3 + 3A(x)[y + B(x)] + \int^x 3a(\tau)B(\tau)d\tau \right\},$

$$a(x) = -\frac{b(x)}{3[A(x) + B(x)^2]},$$

(ii) $\quad \mu(x,y) = \left\{ \dfrac{2y + B(x) + \alpha B(x)^{-1} + \beta B(x)}{2y + B(x) + \alpha B(x)^{-1} - \beta B(x)} \right\}^{1/\beta},$

$$a(x) = \frac{b(x)}{4B(x)} \left\{ [B(x) + \alpha B(x)^{-1}]^2 - \beta^2 B(x)^2 \right\},$$

where α and β denote arbitrary constants.

Chapter Two
One-parameter groups and Lie series

2.1 Introduction

In this chapter we introduce the concepts of one-parameter groups and Lie series. For one-parameter groups there are two important results. Firstly the method of obtaining the *global form* of the group from the *infinitesimal form*. Secondly the existence of *canonical coordinates* for the group. For Lie series the important and remarkable result is the so-called *Commutation theorem*. These concepts are discussed below.

2.2 One-parameter transformation groups

In the (x, y) plane we say that the transformation

$$x_1 = f(x, y, \epsilon), \quad y_1 = g(x, y, \epsilon), \tag{2.1}$$

is a *one-parameter group of transformations* if the following properties hold:

(i) (identity) the value $\epsilon = 0$ characterizes the identity transformation,

$$x = f(x, y, 0), \quad y = g(x, y, 0).$$

(ii) (inverse) the parameter $-\epsilon$ characterizes the inverse transformation,

$$x = f(x_1, y_1, -\epsilon), \quad y = g(x_1, y_1, -\epsilon).$$

(iii) (closure) if $x_2 = f(x_1, y_1, \delta), \quad y_2 = g(x_1, y_1, \delta)$ then the product of the two transformations is also a member of the set of transformations (2.1) and moreover is characterized by the parameter $\epsilon + \delta$, that is

$$x_2 = f(x, y, \epsilon + \delta), \quad y_2 = g(x, y, \epsilon + \delta).$$

Again we remark that the usual associativity law for groups follows from the closure property. Some simple examples of one-parameter groups are:

(a) $\qquad x_1 = x, \quad y_1 = y + \epsilon$ (translational group),

(b) $\qquad x_1 = e^\epsilon x, \quad y_1 = e^\epsilon y$ (stretching group),

(c) $\qquad x_1 = x \cos\epsilon - y \sin\epsilon, \quad y_1 = x \sin\epsilon + y \cos\epsilon$ (rotation group).

Example 2.1 Show that the rotation group (c) does indeed form a one-parameter group of transformations as defined above.

In order to show (c) forms a group we have immediately $x_1 = x$ and $y_1 = y$ when $\epsilon = 0$ so (i) is satisfied. On inverting we obtain

$$x = x_1 \cos \epsilon + y_1 \sin \epsilon, \quad y = y_1 \cos \epsilon - x_1 \sin \epsilon,$$

so that $-\epsilon$ characterizes the inverse and (ii) is satisfied. For (iii) we see that if $x_2 = x_1 \cos \delta - y_1 \sin \delta$ and $y_2 = x_1 \sin \delta + y_1 \cos \delta$ then we have

$$x_2 = (x \cos \epsilon - y \sin \epsilon) \cos \delta - (x \sin \epsilon + y \cos \epsilon) \sin \delta$$
$$= x \cos(\epsilon + \delta) - y \sin(\epsilon + \delta),$$

and

$$y_2 = (x \cos \epsilon - y \sin \epsilon) \sin \delta + (x \sin \epsilon + y \cos \epsilon) \cos \delta$$
$$= x \sin(\epsilon + \delta) + y \cos(\epsilon + \delta),$$

and therefore (iii) is satisfied.

The functions $f(x, y, \epsilon)$ and $g(x, y, \epsilon)$ are referred to as the *global form* of the group. If for small values of the parameter ϵ we expand (2.1) then since $\epsilon = 0$ gives the identity we have

$$x_1 = x + \epsilon \left(\frac{dx_1}{d\epsilon}\right)_{\epsilon=0} + \mathbf{O}(\epsilon^2), \quad y_1 = y + \epsilon \left(\frac{dy_1}{d\epsilon}\right)_{\epsilon=0} + \mathbf{O}(\epsilon^2), \quad (2.2)$$

where $\mathbf{O}(\epsilon^2)$ indicates terms involving only ϵ^2 and higher powers of ϵ. If we introduce functions $\xi(x, y)$ and $\eta(x, y)$ by

$$\left(\frac{dx_1}{d\epsilon}\right)_{\epsilon=0} = \xi(x, y), \quad \left(\frac{dy_1}{d\epsilon}\right)_{\epsilon=0} = \eta(x, y), \quad (2.3)$$

then we obtain

$$x_1 = x + \epsilon \xi(x, y) + \mathbf{O}(\epsilon^2), \quad y_1 = y + \epsilon \eta(x, y) + \mathbf{O}(\epsilon^2), \quad (2.4)$$

and (2.4) is referred to as the *infinitesimal form* of the group. *The crucial property of one-parameter transformation groups is that given the infinitesimal form of the group we can deduce the global form by integrating the following autonomous system of differential equations,*

$$\frac{dx_1}{d\epsilon} = \xi(x_1, y_1), \quad \frac{dy_1}{d\epsilon} = \eta(x_1, y_1), \quad (2.5)$$

subject to the initial conditions,

$$x_1 = x, \quad y_1 = y, \text{ when } \epsilon = 0. \quad (2.6)$$

A proof of this result can be found in Dickson (1924)(page 293). Here we merely indicate its validity by means of a simple example.

Example 2.2 Derive the infinitesimal form of the rotation group (c) and by integration of the autonomous system (2.5) deduce the global form of the group.

For the rotation group (c) we have

$$\frac{dx_1}{d\epsilon} = -x\sin\epsilon - y\cos\epsilon, \quad \frac{dy_1}{d\epsilon} = x\cos\epsilon - y\sin\epsilon,$$

and therefore on setting $\epsilon = 0$ we have from (2.3)

$$\xi(x,y) = -y, \quad \eta(x,y) = x.$$

Thus in this case we need to integrate

$$\frac{dx_1}{d\epsilon} = -y_1, \quad \frac{dy_1}{d\epsilon} = x_1,$$

subject to the initial conditions (2.6). Introducing the complex variables $z = x + iy$ and $z_1 = x_1 + iy_1$ we obtain

$$\frac{dz_1}{d\epsilon} = iz_1,$$

and thus,

$$\log z_1 = i\epsilon + \log z,$$

where we have used the initial conditions (2.6). On equating real and imaginary parts of $z_1 = e^{i\epsilon}z$ we can readily deduce the global form of the rotation group (c). If we introduce polar coordinates (r, θ) defined by

$$r = (x^2 + y^2)^{1/2}, \quad \theta = \tan^{-1}(y/x),$$

then we have $z = re^{i\theta}$ and from $z_1 = e^{i\epsilon}z$ we see that the global form of the rotation group (c) can be written alternatively as $r_1 = r$ and $\theta_1 = \theta + \epsilon$. That is, in terms of (r, θ) coordinates the rotation group has the appearance of the translation group. This is a general property of one-parameter transformation groups.

For any given one-parameter transformation group (2.1) there exists functions $u(x,y)$ and $v(x,y)$ such that the global form of the group becomes

$$u(x_1, y_1) = u(x, y), \quad v(x_1, y_1) = v(x, y) + \epsilon. \tag{2.7}$$

The function $u(x, y)$ is said to be an *invariant* of the group while together (u, v) are referred to as the *canonical coordinates* of the group.

(i) <u>Methods for finding $u(x,y)$.</u> From (2.5) we obtain

$$\frac{dx_1}{dy_1} = \frac{\xi(x_1,y_1)}{\eta(x_1,y_1)},\tag{2.8}$$

which we suppose integrates to yield $u(x_1,y_1)$ = constant so that from the initial conditions (2.6) we deduce the first equation of (2.7) and $u(x,y)$ is known. Alternatively $u(x,y)$ may be deduced directly from (2.1) simply by eliminating ϵ from (2.1)$_1$ and (2.1)$_2$. We note that if $u(x,y)$ is an invariant then so also is any function of $u(x,y)$, namely $\phi(u)$.

(ii) <u>Method for finding $v(x,y)$.</u> In the integration of (2.8) let $\alpha = u(x,y)$ and suppose that from $u(x_1,y_1) = \alpha$ we can deduce the explicit relation $y_1 = \phi(x_1,\alpha)$. Now for the purposes of integration in (2.5), α is a constant and from (2.5)$_1$ we have

$$\frac{dx_1}{d\epsilon} = \xi[x_1, \phi(x_1,\alpha)].\tag{2.9}$$

If for some constant x_0 we define $\psi(x,\alpha)$ by

$$\psi(x,\alpha) = \int_{x_0}^{x} \frac{dt}{\xi[t,\phi(t,\alpha)]},\tag{2.10}$$

then from (2.6) and (2.9) we can deduce (2.7)$_2$ where $v(x,y) = \psi[x,u(x,y)]$ and hence $v(x,y)$ is known.

Example 2.3 Show that the transformation

$$x_1 = \frac{x}{(1+\epsilon x)}, \quad y_1 = (1+\epsilon x)^2 y,\tag{2.11}$$

is a one-parameter group and find the canonical coordinates (u,v).

The reader can verify that (i), (ii) and (iii) of the definition of a one-parameter are indeed satisfied. Now for small values of ϵ we have

$$x_1 = x - \epsilon x^2 + O(\epsilon^2), \quad y_1 = y + 2\epsilon xy + O(\epsilon^2),$$

so that from (2.4) $\xi(x,y) = -x^2$ and $\eta(x,y) = 2xy$. Alternatively on differentiating (2.11) with respect to ϵ we have

$$\frac{dx_1}{d\epsilon} = -\frac{x^2}{(1+\epsilon x)^2} = -x_1^2, \quad \frac{dy_1}{d\epsilon} = 2(1+\epsilon x)xy = 2x_1 y_1,\tag{2.12}$$

and (2.5) confirms these expressions for $\xi(x,y)$ and $\eta(x,y)$. From (2.12) we have

$$\frac{dx_1}{dy_1} = -\frac{x_1}{2y_1},$$

One-parameter Groups and Lie Series

which on integrating gives $u(x,y) = x^2 y$ as an invariant while from $(2.12)_1$ we see that $v(x,y) = x^{-1}$ satisfies (2.7). Alternatively the invariant $x^2 y$ could be deduced directly from (2.11) by eliminating ϵ.

Example 2.4 Show that ξ and η are related to the canonical coordinates u and v by the relations,

$$\xi(x,y) = -\frac{\partial u}{\partial y} \bigg/ \frac{\partial(u,v)}{\partial(x,y)}, \quad \eta(x,y) = \frac{\partial u}{\partial x} \bigg/ \frac{\partial(u,v)}{\partial(x,y)}, \tag{2.13}$$

where the Jacobian is given by

$$\frac{\partial(u,v)}{\partial(x,y)} = \frac{\partial u}{\partial x}\frac{\partial v}{\partial y} - \frac{\partial u}{\partial y}\frac{\partial v}{\partial x}.$$

On differentiating (2.7) with respect to ϵ we obtain

$$\frac{\partial u_1}{\partial x_1}\frac{dx_1}{d\epsilon} + \frac{\partial u_1}{\partial y_1}\frac{dy_1}{d\epsilon} = 0, \quad \frac{\partial v_1}{\partial x_1}\frac{dx_1}{d\epsilon} + \frac{\partial v_1}{\partial y_1}\frac{dy_1}{d\epsilon} = 1, \tag{2.14}$$

where u_1 and v_1 denote $u(x_1, y_1)$ and $v(x_1, y_1)$ respectively. From (2.5) and (2.14) we have on replacing (x_1, y_1) by (x, y),

$$\frac{\partial u}{\partial x}\xi + \frac{\partial u}{\partial y}\eta = 0, \quad \frac{\partial v}{\partial x}\xi + \frac{\partial v}{\partial y}\eta = 1,$$

and (2.13) can be deduced immediately from these relations.

Example 2.5 A transformation in the (x,y) plane is area preserving if

$$\frac{\partial(x_1, y_1)}{\partial(x,y)} = 1. \tag{2.15}$$

Show that (2.1) is area preserving if and only if

$$\frac{\partial(u,v)}{\partial(x,y)} = \Phi(u),$$

where Φ is a function of u only.

From (2.4) and (2.15) we can deduce on equating terms of order ϵ,

$$\frac{\partial \xi}{\partial x} + \frac{\partial \eta}{\partial y} = 0. \tag{2.16}$$

From (2.13) and (2.16) we can deduce

$$\frac{\partial [\partial(u,v)/\partial(x,y), u]}{\partial(x,y)} = 0,$$

and the required condition follows.

[**Aside** It may be of interest to note that since the set of area preserving transformations forms a group, the infinitesimal condition (2.16) is precisely the same as the global condition. That is, if we differentiate (2.15) with respect to ϵ we have

$$\frac{\partial(dx_1/d\epsilon, y_1)}{\partial(x,y)} + \frac{\partial(x_1, dy_1/d\epsilon)}{\partial(x,y)} = 0,$$

and on multiplying this equation by $\dfrac{\partial(x,y)}{\partial(x_1,y_1)}$ we obtain

$$\frac{\partial(dx_1/d\epsilon, y_1)}{\partial(x_1,y_1)} + \frac{\partial(x_1, dy_1/d\epsilon)}{\partial(x_1,y_1)} = 0,$$

so that we have

$$\frac{\partial}{\partial x_1}\left(\frac{dx_1}{d\epsilon}\right) + \frac{\partial}{\partial y_1}\left(\frac{dy_1}{d\epsilon}\right) = 0. \tag{2.17}$$

On using (2.5) we see that (2.17) is the same condition as (2.16).]

2.3 Lie series and the Commutation theorem

Suppose we have the group (2.1) with infinitesimal version (2.4). We define the differential operator L by

$$L = \xi(x,y)\frac{\partial}{\partial x} + \eta(x,y)\frac{\partial}{\partial y}. \tag{2.18}$$

Now for any function $\phi(x_1, y_1)$ which does not depend explicitly on ϵ we have

$$\frac{d\phi_1}{d\epsilon} = \frac{\partial \phi_1}{\partial x_1}\frac{dx_1}{d\epsilon} + \frac{\partial \phi_1}{\partial y_1}\frac{dy_1}{d\epsilon}, \tag{2.19}$$

where ϕ_1 denotes $\phi(x_1, y_1)$. From (2.5) and (2.19) we obtain

$$\frac{d\phi_1}{d\epsilon} = L_1(\phi_1), \tag{2.20}$$

One-parameter Groups and Lie Series 23

where L_1 denotes the differential operator L with (x,y) replaced by (x_1, y_1). Similarly we have

$$\frac{d^2\phi_1}{d\epsilon^2} = L_1(L_1(\phi_1)), \quad \frac{d^3\phi_1}{d\epsilon^3} = L_1(L_1(L_1(\phi_1))). \tag{2.21}$$

If we let $\Phi(\epsilon) = \phi(x_1, y_1)$ then by Maclaurin's expansion we have

$$\Phi(\epsilon) = \Phi(0) + \epsilon \left(\frac{d\Phi}{d\epsilon}\right)_{\epsilon=0} + \frac{\epsilon^2}{2!}\left(\frac{d^2\Phi}{d\epsilon^2}\right)_{\epsilon=0} + \frac{\epsilon^3}{3!}\left(\frac{d^3\Phi}{d\epsilon^3}\right)_{\epsilon=0} + \cdots,$$

and thus from (2.20) and (2.21) with $\epsilon = 0$ we obtain

$$\phi(x_1, y_1) = \phi(x, y) + \epsilon L(\phi) + \frac{\epsilon^2}{2!}L^2(\phi) + \frac{\epsilon^3}{3!}L^3(\phi) + \cdots.$$

That is, we have

$$\phi(x_1, y_1) = \sum_{n=0}^{\infty} \frac{\epsilon^n}{n!}L^n(\phi), \tag{2.22}$$

and we refer to such a series as a *Lie series*. We notice that we can write (2.22) as

$$\phi(x_1, y_1) = e^{\epsilon L}\phi(x, y), \tag{2.23}$$

provided we interpret the differential operator $e^{\epsilon L}$ as the series operator,

$$\sum_{n=0}^{\infty} \frac{\epsilon^n}{n!}L^n(\ ).$$

In particular if we take $\phi(x, y)$ to be x and y then form (2.23) we obtain,

$$x_1 = e^{\epsilon L}x, \quad y_1 = e^{\epsilon L}y. \tag{2.24}$$

On combining (2.23) and (2.24) we have the remarkable result,

$$\phi(e^{\epsilon L}x, e^{\epsilon L}y) = e^{\epsilon L}\phi(x, y), \tag{2.25}$$

which is called the *Commutation theorem of Lie series* (see Gröbner and Knapp (1967), page 17).

Example 2.6 For the rotation group (c) deduce the global form of the group by means of (2.24) and the appropriate differential operator L.

In this case the differential operator L is given by

$$L = -y\frac{\partial}{\partial x} + x\frac{\partial}{\partial y},$$

so that $L(x) = -y$ and $L(y) = x$ and the global form of the group can be deduced from (2.24) using the expansions,

$$\cos \epsilon = \sum_{k=0}^{\infty} \frac{(-1)^k \epsilon^{2k}}{(2k)!}, \quad \sin \epsilon = \sum_{k=0}^{\infty} \frac{(-1)^k \epsilon^{2k+1}}{(2k+1)!}.$$

It is worthwhile noting that using Lie series we can give a formal solution of any autonomous system of differential equations given initial values. That is, consider

$$\frac{dX}{dt} = F(X,Y), \quad \frac{dY}{dt} = G(X,Y),$$

and $X = \alpha$, $Y = \beta$ at $t = 0$. The formal solution of this initial value problem is

$$X = e^{tM}\alpha, \quad Y = e^{tM}\beta, \tag{2.26}$$

where the operator M is defined by

$$M = F(\alpha,\beta)\frac{\partial}{\partial \alpha} + G(\alpha,\beta)\frac{\partial}{\partial \beta}.$$

We consider two simple examples.

Example 2.7 Deduce the solution of the single differential equation

$$\frac{dX}{dt} = -X^2,$$

utilizing the above Lie series.

In this case we have

$$M = -\alpha^2 \frac{\partial}{\partial \alpha},$$

and $M(\alpha) = -\alpha^2$, $M^2(\alpha) = 2\alpha^3$ and in general

$$M^n(\alpha) = (-1)^n n! \alpha^{n+1}.$$

One-parameter Groups and Lie Series 25

Hence

$$X = e^{tM}\alpha = \sum_{n=0}^{\infty} \frac{t^n}{n!} M^{(n)}(\alpha) = \alpha \sum_{n=0}^{\infty} (-\alpha t)^n,$$

and thus for $|\alpha t| < 1$ we obtain the solution,

$$X = \frac{\alpha}{1 + \alpha t}.$$

Example 2.8 Using Lie series deduce the solution of

$$\frac{dX}{dt} = AX + BY, \frac{dY}{dt} = -AY + BX, \qquad (2.27)$$

where A and B are constants.

In this case we have

$$M = (A\alpha + B\beta)\frac{\partial}{\partial \alpha} + (B\alpha - A\beta)\frac{\partial}{\partial \beta},$$

so that

$$\begin{aligned}
M(\alpha) &= (A\alpha + B\beta), & M(\beta) &= (B\alpha - A\beta), \\
M^2(\alpha) &= K^2 \alpha, & M^2(\beta) &= K^2 \beta, \\
M^3(\alpha) &= K^2(A\alpha + B\beta), & M^3(\beta) &= K^2(B\alpha - A\beta), \\
M^4(\alpha) &= K^4 \alpha, & M^4(\beta) &= K^4 \beta, \\
M^5(\alpha) &= K^4(A\alpha + B\beta), & M^5(\beta) &= K^4(B\alpha - A\beta),
\end{aligned}$$

and so on, where $K = (A^2 + B^2)^{1/2}$. From these results and (2.26) we can deduce the solutions,

$$\begin{aligned}
2KX &= [K\alpha + (A\alpha + B\beta)]e^{Kt} + [K\alpha - (A\alpha + B\beta)]e^{-Kt}, \\
2KY &= [K\beta + (B\alpha - A\beta)]e^{Kt} + [K\beta - (B\alpha - A\beta)]e^{-Kt},
\end{aligned}$$

which of course could be established by more elementary methods (for example, differential $(2.27)_1$ with respect to t and make use of $(2.27)_2$).

[The following problems which arise in continuum mechanics are useful exercises in manipulating Lie series.]

PROBLEMS

1. In cylindrical polar coordinates (r, θ) a transformation in the plane area preserving if

$$\frac{\partial(r_1, \theta_1)}{\partial(r, \theta)} = \frac{r}{r_1}.$$

Following the note at the end of Example 2.5, differentiate this equation with respect to ϵ and show that,

$$\frac{\partial}{\partial r_1}\left(\frac{dr_1}{d\epsilon}\right) + \frac{1}{r_1}\frac{dr_1}{d\epsilon} + \frac{\partial}{\partial \theta_1}\left(\frac{d\theta_1}{d\epsilon}\right) = 0.$$

Hence deduce there exists a function $\phi(r_1, \theta_1, \epsilon)$ such that

$$\frac{dr_1}{d\epsilon} = -\frac{1}{r_1}\frac{\partial \phi}{\partial \theta_1}, \quad \frac{d\theta_1}{d\epsilon} = \frac{1}{r_1}\frac{\partial \phi}{\partial r_1}.$$

If ϕ does not depend explicitly on ϵ the solution of this system for which $r_1 = r, \theta_1 = \theta$ when $\epsilon = 0$ is a one-parameter group with ϕ as an invariant, that is

$$\phi(r_1, \theta_1) = \phi(r, \theta).$$

2. **Continuation.** In the above problem show that the one-parameter groups corresponding to

(i) $\quad\quad\quad\quad\quad\quad\quad\quad \phi(r, \theta) = Ar^2\theta + B\theta,$

(ii) $\quad\quad\quad\quad\quad\quad\quad \phi(r, \theta) = Ar^2\theta + Br^2 \log r,$

(iii) $\quad\quad\quad\quad\quad\quad\quad \phi(r, \theta) = \frac{A}{4}(\theta + \sin\theta \cos\theta),$

where A and B denote arbitrary constants, are respectively

(i) $\quad r_1 = [e^{-2A\epsilon}r^2 + BA^{-1}(e^{-2A\epsilon} - 1)]^{1/2}, \quad \theta_1 = e^{2A\epsilon}\theta,$

(ii) $\quad r_1 = e^{-A\epsilon}r, \quad \theta_1 = e^{2A\epsilon}\theta + BA^{-1}(e^{2A\epsilon} - 1)\log r + B\epsilon,$

(iii) $\quad r_1 = [r^2 - A\epsilon \cos^2\theta]^{1/2}, \quad \theta_1 = \theta.$

3. Consider,

$$x_1 = \frac{1}{\epsilon}\log(1 + \epsilon x), \quad y_1 = (1 + \epsilon x)y.$$

Calculate $\frac{dx_1}{d\epsilon}, \frac{dy_1}{d\epsilon}$ and express in terms of (x_1, y_1). Hence or otherwise deduce that this is not a one-parameter transformation group.

One-parameter Groups and Lie Series

4. Consider the one-parameter group,

$$x_1 = f(x, y, \epsilon) = x + \epsilon \xi(x, y) + \mathbf{O}(\epsilon^2),$$
$$y_1 = g(x, y, \epsilon) = y + \epsilon \eta(x, y) + \mathbf{O}(\epsilon^2),$$

and introduce the operators,

$$L = \xi \frac{\partial}{\partial x} + \eta \frac{\partial}{\partial y}, \quad P = L + \omega, \quad Q = L - \omega,$$

where ω is defined by,

$$\omega = \frac{\partial \xi}{\partial x} + \frac{\partial \eta}{\partial y}.$$

Let $\phi(x, y)$ and $\psi(x, y)$ denote arbitrary functions and agree to use the notation (ϕ, ψ) for the Jacobian, that is

$$(\phi, \psi) = \frac{\partial(\phi, \psi)}{\partial(x, y)}.$$

We consider the following Lie series,

$$e^{\epsilon L} \phi(x, y) = \sum_{n=0}^{\infty} \frac{\epsilon^n}{n!} L^n(\phi) = \sum_{n=0}^{\infty} \epsilon^n \phi_n(x, y),$$
$$e^{\epsilon L} \psi(x, y) = \sum_{n=0}^{\infty} \frac{\epsilon^n}{n!} L^n(\psi) = \sum_{n=0}^{\infty} \epsilon^n \psi_n(x, y).$$

Show that

(i) $$\phi_n = \frac{1}{n} L(\phi_{n-1}), \quad \psi_n = \frac{1}{n} L(\psi_{n-1}),$$

(ii) $$P(\phi, \psi) = (L(\phi), \psi) + (\phi, L(\psi)).$$

[Hint, for (ii) start by considering $L(\phi, \psi)$.]

5. **Continuation.** If we suppose that

$$(e^{\epsilon L} \phi, e^{\epsilon L} \psi) = \sum_{n=0}^{\infty} \epsilon^n \chi_n(x, y),$$

show that,

(i) $$\chi_n = \sum_{k=0}^{n} (\phi_k, \psi_{n-k}),$$

(ii) $$\chi_n = \frac{1}{n} P(\chi_{n-1}).$$

Hence deduce,

$$(e^{\epsilon L}\phi, e^{\epsilon L}\psi) = \sum_{n=0}^{\infty} \frac{\epsilon^n}{n!} P^n(\phi, \psi) = e^{\epsilon P}(\phi, \psi).$$

[Hint, for (ii) start by considering $P(\chi_{n-1})$ and use (i) and (ii) of previous problem.]

6. **Continuation.** Observe that in particular,

$$(x_1, y_1) = (e^{\epsilon L}x, e^{\epsilon L}y) = e^{\epsilon P}1,$$

and therefore

$$(x_1, y_1) = 1 + \epsilon\omega + \frac{\epsilon^2}{2!}P(\omega) + \frac{\epsilon^3}{3!}P^2(\omega) + \dots.$$

Verify

(i) $$\log(x_1, y_1) = \epsilon\omega + \frac{\epsilon^2}{2!}L(\omega) + \frac{\epsilon^3}{3!}L^2(\omega) + \dots,$$

(ii) $$(x_1, y_1)^{-1} = 1 - \epsilon\omega - \frac{\epsilon^2}{2!}Q(\omega) - \frac{\epsilon^3}{3!}Q^2(\omega) + \dots.$$

7. In cylindrical polar coordinates (r, θ) consider the one-parameter group,

$$r_1 = f(r, \theta, \epsilon) = r + \epsilon\xi(r, \theta) + \mathbf{O}(\epsilon^2),$$
$$\theta_1 = g(r, \theta, \epsilon) = \theta + \epsilon\eta(r, \theta) + \mathbf{O}(\epsilon^2),$$

and introduce operators,

$$L = \xi\frac{\partial}{\partial r} + \eta\frac{\partial}{\partial \theta}, \quad P_1 = L + \omega_1, \quad P_2 = L + \omega_2,$$

where ω_1 and ω_2 are defined by

$$\omega_1 = \frac{\partial\xi}{\partial r} + \frac{\partial\eta}{\partial\theta}, \quad \omega_2 = \frac{\xi}{r}.$$

Suppose that

$$r_1 = e^{\epsilon L}r = \sum_{n=0}^{\infty} \frac{\epsilon^n}{n!} L^n(r) = \sum_{n=0}^{\infty} \epsilon^n f_n(r, \theta),$$

$$\lambda = \frac{r_1}{r} = \sum_{n=0}^{\infty} \epsilon^n \lambda_n(r, \theta),$$

$$\mu = \frac{\partial(r_1, \theta_1)}{\partial(r, \theta)} = \sum_{n=0}^{\infty} \epsilon^n \mu_n(r, \theta).$$

Verify

(i) $$f_n = \frac{1}{n} L(f_{n-1}),$$
(ii) $$\lambda_n = \frac{1}{n} P_2(\lambda_{n-1}),$$
(iii) $$\lambda = e^{\epsilon P_2} 1.$$

8. **Continuation.** Observe from Problem 6,

$$\mu = e^{\epsilon P_1} 1.$$

Suppose that,

$$\lambda\mu = \frac{r_1}{r} \frac{\partial(r_1, \theta_1)}{\partial(r, \theta)} = \sum_{n=0}^{\infty} \epsilon^n \sigma_n(r, \theta).$$

Verify,

(i) $$\sigma_n = \sum_{k=0}^{n} \lambda_k \mu_{n-k},$$
(ii) $$\sigma_n = \frac{1}{n} P_3(\sigma_{n-1}),$$

where P_3 is given by

$$P_3 = L + (\omega_1 + \omega_2).$$

Hence conclude,

$$\frac{r_1}{r} \frac{\partial(r_1, \theta_1)}{\partial(r, \theta)} = e^{\epsilon P_2} 1 e^{\epsilon P_1} 1 = e^{\epsilon P_3} 1.$$

[Hint, for (ii) start by considering $L(\sigma_{n-1})$.]

Chapter Three
Invariance of standard linear ordinary differential equations

3.1 Introduction

It is well known that linear differential equations for $y(x)$ remain linear under transformations of the form,
$$x_1 = f(x, \epsilon), \quad y_1 = g(x, \epsilon)y. \tag{3.1}$$

Throughout this chapter we consider only transformations (3.1) which we suppose form a one-parameter group such that infinitesimally we have

$$x_1 = x + \epsilon \xi(x) + \mathbf{O}(\epsilon^2), \quad y_1 = y + \epsilon \eta(x)y + \mathbf{O}(\epsilon^2). \tag{3.2}$$

We look for groups (3.1) which leave standard linear equations invariant and deduce the form of the differential equation in terms of canonical coordinates (u, v) (see (2.7)). Initially the reader may well consider this approach irrelevant for such equations and of course the astute reader will see that for first order equations we still end up solving an equation by classical methods which is comparable in difficulty to the original one. The object of the exercise being to demonstrate the group approach in familiar situations with a view to the student obtaining some insight into the relation between solutions and groups leaving the equation invariant. Moreover even linear equations are not always readily solved and the results obtained in Sections 3.4 and 3.5 by this method are non-trivial and appear not to have been given elsewhere.

3.2 First order equation $y' + p(x)y = q(x)$

For the first order differential equations our primary objective is to introduce new variables such that the equation becomes separable. For equations invariant under a one-parameter group of transformations the appropriate new variables are the canonical coordinates (u, v) of the group. We illustrate this procedure with the standard first order equation,

$$\frac{dy}{dx} + p(x)y = q(x). \tag{3.3}$$

For convenience we introduce the function $s(x)$ by

$$s(x) = e^{\int_{x_0}^{x} p(t)dt}, \qquad (3.4)$$

where x_0 is some constant. With this definition the solution of (3.3) is known to be given by

$$s(x)y - \int_{x_0}^{x} s(t)q(t)dt = C, \qquad (3.5)$$

where C is a constant.

We now deduce (3.5) by finding a group (3.1) which leaves (3.3) invariant, that is

$$\frac{dy_1}{dx_1} + p(x_1)y_1 = q(x_1).$$

From this equation and (3.1) we deduce

$$\frac{dy}{dx} + \left\{ \frac{g'(x)}{g(x)} + f'(x)p(f) \right\} y = \frac{f'(x)}{g(x)} q(f),$$

which becomes (3.3) provided $f(x)$ and $g(x)$ are such that

$$p(x) = \frac{g'(x)}{g(x)} + f'(x)p(f), \qquad q(x) = \frac{f'(x)}{g(x)} q(f).$$

From these equations and

$$f(x) = x + \epsilon\xi(x) + \mathbf{O}(\epsilon^2), \qquad g(x) = 1 + \epsilon\eta(x) + \mathbf{O}(\epsilon^2), \qquad (3.6)$$
$$p(f) = p(x) + \epsilon\xi(x)p'(x) + \mathbf{O}(\epsilon^2), \qquad q(f) = q(x) + \epsilon\xi(x)q'(x) + \mathbf{O}(\epsilon^2),$$

we obtain on equating terms of order ϵ,

$$\eta' + \xi'p + \xi p' = 0, \qquad \xi'q + \xi q' - \eta q = 0.$$

Hence we have

$$\eta + \xi p = C_1, \qquad \xi' + \xi\left(p + \frac{q'}{q}\right) = C_1, \qquad (3.7)$$

where C_1 is a constant. Thus

$$\xi(x) = \frac{1}{q(x)s(x)} \left\{ C_1 \int_{x_0}^{x} s(t)q(t)dt + C_2 \right\},$$
$$\qquad (3.8)$$
$$\eta(x) = C_1 - p(x)\xi(x),$$

Invariance of Standard Linear Ordinary Differential Equations 33

where C_2 is a further constant. [Of course in obtaining these results we have had to solve an equation of the type (3.3), namely (3.7)$_2$.] Now the global form of the group (3.1) is obtained by integrating (see (2.5) and (2.6))

$$\frac{dx_1}{d\epsilon} = \xi(x_1), \quad \frac{dy_1}{d\epsilon} = \eta(x_1)y_1, \tag{3.9}$$

subject to the intital conditions $x_1 = x$, $y_1 = y$ when $\epsilon = 0$. From (3.8)$_2$ and (3.9) we have

$$\frac{1}{y_1}\frac{dy_1}{d\epsilon} = C_1 - p(x_1)\frac{dx_1}{d\epsilon},$$

and therefore, by integrating this equation we obtain,

$$y_1 s(x_1) = e^{C_1 \epsilon} y s(x), \tag{3.10}$$

where $s(x)$ is defined by (3.4). From (3.8)$_1$ and (3.9)$_1$ we have

$$\frac{q(x_1)s(x_1)dx_1}{\left\{C_1 \int_{x_0}^{x_1} s(t)q(t)dt + C_2\right\}} = d\epsilon, \tag{3.11}$$

and there are two cases to be considered.

Firstly if $C_1 \neq 0$ then (3.11) gives

$$\frac{1}{C_1}\log\left\{C_1 \int_{x_0}^{x_1} s(t)q(t)dt + C_2\right\} = \frac{1}{C_1}\log\left\{C_1 \int_{x_0}^{x} s(t)q(t)dt + C_2\right\} + \epsilon,$$

and from this equation and (3.10) we can deduce that our canonical coordinates (u, v) (see (2.7)) are given by

$$u(x, y) = \frac{s(x)y}{\left\{C_1 \int_{x_0}^{x} s(t)q(t)dt + C_2\right\}},$$

$$v(x, y) = \frac{1}{C_1}\log\left\{C_1 \int_{x_0}^{x} s(t)q(t)dt + C_2\right\}.$$

In these coordinates it can be readily verified that the differential equation (3.3) becomes

$$\frac{du}{dv} = 1 - C_1 u,$$

which is separable and integrates to give

$$-\frac{1}{C_1}\log(1 - C_1 u) = v - \frac{1}{C_1}\log C_3,$$

that is,
$$(1 - C_1 u)e^{C_1 v} = C_3,$$

where C_3 is a constant. This equation can be reconciled with (3.5) where the arbitrary constant C in (3.5) is $(C_2 - C_3)/C_1$.

Secondly if $C_1 = 0$ then from (3.10) and (3.11) we have canonical coordinates
$$u(x,y) = s(x)y, \quad v(x,y) = \frac{1}{C_2}\int_{x_0}^{x} s(t)q(t)dt,$$

and in these coordinates (3.3) becomes
$$\frac{du}{dv} = C_2.$$

Again our equation in canonical coordinates is separable and can be integrated to give
$$u - C_2 v = C,$$

which can also be reconciled with (3.5) where the constant C is the same in both equations. We remark that (3.3) is invariant under other groups, in addition to those considered here.

3.3 Second order homogeneous equation $y'' + p(x)y = 0$

For second order linear homogeneous equations we can without loss of generality (see Problem 8) consider the normal form of the equation, namely
$$\frac{d^2 y}{dx^2} + p(x)y = 0. \tag{3.12}$$

We shall assume $\phi_1(x)$ and $\phi_2(x)$ are two linearly independent solutions of (3.12) and for convenience we suppose their Wronskian is unity, that is
$$\phi_1 \phi_2' - \phi_2 \phi_1' = 1. \tag{3.13}$$

Our objective here is to relate ϕ_1 and ϕ_2 to a one-parameter group (3.1) which leaves (3.12) invariant. For higher order linear differential equations the use of canonical coordinates (u, v) simplifies the equation to one with constant coefficients.

From (3.1) we have

$$\frac{dy_1}{dx_1} = \frac{g}{f'}\frac{dy}{dx} + \frac{g'}{f'}y,$$

and

$$\frac{d^2y_1}{dx_1^2} = \frac{g}{f'^2}\frac{d^2y}{dx^2} + \left(\frac{2g'}{f'^2} - \frac{gf''}{f'^3}\right)\frac{dy}{dx} + \left(\frac{g''}{f'^2} - \frac{g'f''}{f'^3}\right)y. \tag{3.14}$$

Clearly if (3.12) is to remain invariant there can be no term involving y'. On equating the coefficient of y' to zero in (3.14) we obtain $f'(x)/g(x)^2$ constant, which must be unity if (3.1) is a one-parameter group and therefore we have

$$f'(x) = g(x)^2. \tag{3.15}$$

From (3.14) and (3.15) we find that the differential equation

$$\frac{d^2y_1}{dx_1^2} + p(x_1)y_1 = 0,$$

becomes

$$\frac{d^2y}{dx^2} + \left(\frac{g''}{g} - 2\frac{g'^2}{g^2} + p(f)g^4\right)y = 0,$$

and thus (3.12) remains invariant provided

$$\frac{g''}{g} - 2\frac{g'^2}{g^2} + p(f)g^4 = p(x). \tag{3.16}$$

Equation (3.15) and (3.16) constitute two equations for the determination of the group (3.1). From (3.6), (3.15) and (3.16) we find on equating terms of order ϵ,

$$\xi' = 2\eta, \quad \frac{\xi'''}{2} + 2p\xi' + p'\xi = 0. \tag{3.17}$$

The equation (3.17)$_2$ for $\xi(x)$ is a formally self-adjoint third order differential equation (sometimes called anti self-adjoint, see Murphy (1960), page 199) with first integral

$$\frac{1}{4}(2\xi\xi'' - \xi'^2) + p\xi^2 = \text{constant}, \tag{3.18}$$

which can be verified by differentiation. It is well known (see either Problem 20 or Murphy (1960), page 200) that the general solution of (3.17)$_2$ is given by

$$\xi = A\phi_1^2 + 2B\phi_1\phi_2 + C\phi_2^2, \tag{3.19}$$

where A, B and C denote arbitrary constants. [In order to see that (3.19) is the general solution consider for example $\hat{\xi} = \phi_1 \phi_2$ then we have

$$\hat{\xi}' = \phi_1 \phi_2' + \phi_2 \phi_1', \quad \hat{\xi}'' = \phi_1 \phi_2'' + 2\phi_1' \phi_2' + \phi_2 \phi_1''.$$

Observe that from the original differential equation we have

$$\hat{\xi}'' = 2(\phi_1' \phi_2' - p\phi_1 \phi_2),$$

and substitution of this expression and those for $\hat{\xi}'$ and $\hat{\xi}$ into $(3.17)_2$ gives zero.] Further from (3.12) and (3.19) we can deduce

$$\xi' = 2[A\phi_1 \phi_1' + B(\phi_1 \phi_2' + \phi_2 \phi_1') + C\phi_2 \phi_2'],$$
$$\xi'' = 2(A\phi_1'^2 + 2B\phi_1' \phi_2' + C\phi_2'^2) - 2p\xi,$$

and on substitution into (3.18) we find on using (3.13) that (3.18) becomes

$$\frac{1}{4}(2\xi\xi'' - \xi'^2) + p\xi^2 = (AC - B^2). \tag{3.20}$$

Now the global form of the one-parameter group (3.1) is obtained by solving the differential equations

$$\frac{dx_1}{d\epsilon} = \xi(x_1), \quad \frac{dy_1}{d\epsilon} = \frac{\xi'(x_1)}{2} y_1,$$

subject to the initial conditions $x_1 = x$, $y_1 = y$ and $\epsilon = 0$. We find that suitable canonical coordinates (u, v) are given by

$$u(x, y) = \frac{y}{\xi(x)^{1/2}}, \quad v(x, y) = \int_{x_0}^{x} \frac{dt}{\xi(t)},$$

where x_0 is some constant. In terms of (u, v) we find that the differential equation (3.12) becomes

$$\frac{d^2 u}{dv^2} + \left\{ \frac{1}{4}(2\xi\xi'' - \xi'^2) + p\xi^2 \right\} u = 0.$$

But from (3.20) we see that the differential equation finally becomes

$$\frac{d^2 u}{dv^2} + (AC - B^2)u = 0. \tag{3.21}$$

Thus for example, if $(AC - B^2)$ is positive the general solution of (3.12) is given by

$$y(x) = \xi(x)^{1/2} \left\{ C_1 \cos \left(K \int_{x_0}^{x} \frac{dt}{\xi(t)} \right) + C_2 \sin \left(K \int_{x_0}^{x} \frac{dt}{\xi(t)} \right) \right\}, \tag{3.22}$$

Invariance of Standard Linear Ordinary Differential Equations 37

where C_1, C_2 are arbitrary constants and $K = (AC - B^2)^{1/2}$. We have therefore established the relationship between the general solution of (3.12) and the infinitesimal version of the one-parameter group of transformations leaving (3.12) invariant. In a sense, (3.22) is the 'inverse' of (3.19). From (3.19) we see that if we know a group leaving (3.12) unaltered then essentially we know a quadratic relation between the linearly independent solutions of (3.12). We remark that solutions of (3.12) in the form of (3.22) have been given previously although the function $\xi(x)$ appearing in (3.22) has not been identified with the one-parameter group leaving the differential equation invariant (see for example Coppel (1971), page 19).

Example 3.1 Illustrate the above theory with reference to the simple Euler equation,

$$\frac{d^2y}{dx^2} + \frac{y}{4x^2} = 0.$$

In this case the equation is clearly invariant under the group $x_1 = e^\epsilon x$ and $y_1 = y$ so that $\xi(x) = x$. Since we have linearly independent solutions $\phi_1(x) = x^{1/2}$ and $\phi_2(x) = x^{1/2} \log x$ we see from (3.19) $A = 1$, $B = C = 0$ and therefore from (3.21) we have $u = C_1 v + C_2$. This expression confirms our linearly independent solutions since in this case $u = y/x^{1/2}$ and $v = \log x$.

3.4 Third order homogeneous equation $y''' + p(x)y' + q(x)y = 0$

We see from Problem 12 that for third order linear homogeneous differential equations we can without loss of generality consider the equation

$$\frac{d^3y}{dx^3} + p(x)\frac{dy}{dx} + q(x)y = 0. \tag{3.23}$$

In this section we suppose that $\phi_1(x)$ and $\phi_2(x)$ are linearly independent solutions of the second order equation

$$\frac{d^2y}{dx^2} + \frac{p(x)}{4}y = 0, \tag{3.24}$$

such that their Wronskian is unity. We are concerned with finding one-parameter groups of the form (3.1) which leave (3.23) invariant.

From a further differentiation of (3.14) we obtain

$$\frac{d^3y_1}{dx_1^3} = \frac{g}{f'^3}\frac{d^3y}{dx^3} + 3\left(\frac{g'}{f'^3} - \frac{gf''}{f'^4}\right)\frac{d^2y}{dx^2} + \left(\frac{3g''}{f'^3} - \frac{6g'f''}{f'^4} + \frac{3gf''^2}{f'^5} - \frac{gf'''}{f'^4}\right)\frac{dy}{dx}$$

$$+ \left(\frac{g'''}{f'^3} + \frac{3g'f''^2}{f'^5} - \frac{3g''f''}{f'^4} - \frac{g'f'''}{f'^4}\right)y. \tag{3.25}$$

If (3.23) is to remain invariant we require the coefficient of y'' in (3.25) to be zero. From this condition we deduce

$$f'(x) = g(x), \tag{3.26}$$

and (3.25) becomes

$$\frac{d^3 y_1}{dx_1^3} = \frac{1}{g^2}\frac{d^3 y}{dx^3} + \left(\frac{2g''}{g^3} - \frac{3g'^2}{g^4}\right)\frac{dy}{dx} + \left(\frac{g'''}{g^3} + \frac{3g'^3}{g^5} - \frac{4g'g''}{g^4}\right)y.$$

Using this equation and

$$\frac{d^3 y_1}{dx_1^3} + p(x_1)\frac{dy_1}{dx_1} + q(x_1)y_1 = 0,$$

we obtain on multiplying by g^2 the equation,

$$\frac{d^3 y}{dx^3} + \left(\frac{2g''}{g} - \frac{3g'^2}{g^2} + p(f)g^2\right)\frac{dy}{dx} + \left(\frac{g'''}{g} + \frac{3g'^3}{g^3} - \frac{4g'g''}{g^2} + p(f)gg' + q(f)g^3\right)y = 0.$$

For invariance this equation must be identical with (3.23) and therefore $f(x)$ and $g(x)$ as well as satisfying (3.26) must also satisfy

$$\frac{2g''}{g} - \frac{3g'^2}{g^2} + p(f)g^2 = p(x),$$

$$\frac{g'''}{g} + \frac{3g'^3}{g^3} - \frac{4g'g''}{g^2} + p(f)gg' + q(f)g^3 = q(x). \tag{3.27}$$

From (3.6) and (3.26) we obtain $\eta = \xi'$ while from (3.27) we have

$$2\xi''' + 2p\xi' + p'\xi = 0,$$

$$\xi'''' + p\xi'' + 3q\xi' + q'\xi = 0, \tag{3.28}$$

and these equations are only consistent if

$$\xi^3\left(q - \frac{p'}{2}\right) = D, \tag{3.29}$$

where D is a constant. Clearly if (3.23) is self-adjoint (see Problem 14) then $p' = 2q$ and (3.29) is trivially satisfied with the constant D zero. However if (3.23) is not self-adjoint then $\xi(x)$ must be both a solution of $(3.28)_1$ as well as satisfying (3.29). In terms of solutions of (3.24) the general solution of $(3.28)_1$ is given by

$$\xi = A\phi_1^2 + 2B\phi_1\phi_2 + C\phi_2^2, \tag{3.30}$$

and we have

$$2\xi\xi'' - \xi'^2 + p\xi^2 = 4(AC - B^2), \tag{3.31}$$

where A, B and C denote arbitrary constants and we have used the fact that the Wronskian of ϕ_1 and ϕ_2 is unity. If (3.23) is not self-adjoint then for a given $p(x)$ we need to assume $q(x)$ is given by

$$q(x) = \frac{1}{2}\frac{dp}{dx} + \frac{D}{\xi(x)^3}, \tag{3.32}$$

where $\xi(x)$ is given by (3.30).

From the equations

$$\frac{dx_1}{d\epsilon} = \xi(x_1), \quad \frac{dy_1}{d\epsilon} = \xi'(x_1)y_1,$$

we find that suitable canonical coordinates (u, v) are given by

$$u(x, y) = \frac{y}{\xi(x)}, \quad v(x, y) = \int_{x_0}^{x} \frac{dt}{\xi(t)},$$

for some constant x_0. In these coordinates the differential equation (3.23) can be shown to become

$$\frac{d^3u}{dv^3} + (2\xi\xi'' - \xi'^2 + p\xi^2)\frac{du}{dv} + \xi^2(\xi''' + p\xi' + q\xi)u = 0,$$

which on using $(3.28)_1$, (3.29) and (3.31) finally becomes

$$\frac{d^3u}{dv^3} + 4(AC - B^2)\frac{du}{dv} + Du = 0. \tag{3.33}$$

Thus for third order differential equations (3.23) which are not self-adjoint, we can for a given function $p(x)$ obtain a one-parameter group (3.1) which leaves the equation invariant provided the function $q(x)$ has the form (3.32) for suitable constants A, B, C and D. If this is the case then the solution of (3.23) reduces to solving the third order linear equation (3.33) with constant coefficients. If (3.23) happens to be self-adjoint then the general solution can be obtained in the usual way for such equations from the solutions of (3.24) (Murphy (1960), page 200).

Example 3.2 Illustrate the above theory for the case when $\phi_1(x)$ and $\phi_2(x)$ are linearly independent solutions of the Euler equation given in Example 3.1.

In this case $p(x) = x^{-2}$, $\phi_1(x) = x^{1/2}$, $\phi_2(x) = x^{1/2} \log x$ and from (3.32) $q(x)$ must be given by

$$q(x) = x^{-3}\{[A + 2B \log x + C(\log x)^2]^{-3}D - 1\},$$

for some constants A, B, C and D. If this is the case (3.33) can be solved by an expression of the form

$$u = C_1 e^{k_1 v} + C_2 e^{k_2 v} + C_3 e^{k_3 v},$$

where C_1, C_2 and C_3 denote three arbitrary constants and k_1, k_2 and k_3 are the three roots of the cubic equation

$$k^3 + 4(AC - B^2)k + D = 0.$$

Assuming that the roots of this cubic are readily identified we may proceed to deduce three linearly independent solutions of the original equation of the form (3.23).

3.5 Fourth order self adjoint equation $y'''' + [p(x)y']' + q(x)y = 0$

From Problems 16 and 17 we deduce that the general fourth order self-adjoint equation can be taken as

$$\frac{d^4y}{dx^4} + \frac{d}{dx}\left(p(x)\frac{dy}{dx}\right) + q(x)y = 0. \tag{3.34}$$

In this section we suppose $\phi_1(x)$ and $\phi_2(x)$ are linearly independent solutions of

$$\frac{d^2y}{dx^2} + \frac{p(x)}{10}y = 0, \tag{3.35}$$

such that their Wronskian is unity.

On a further differentiation of (3.25) we find that the coefficient of y''' is zero provided,

$$f'(x) = g(x)^{2/3}, \tag{3.36}$$

in which case we obtain,

$$\frac{d^4 y_1}{dx_1^4} = \frac{1}{g^{5/3}}\left\{\frac{d^4y}{dx^4} + \frac{10}{3}\left(\frac{g''}{g} - \frac{4}{3}\frac{g'^2}{g^2}\right)\frac{d^2y}{dx^2} + \frac{10}{3}\left(\frac{g'''}{g} - \frac{11}{3}\frac{g'g''}{g^2} + \frac{8}{3}\frac{g'^3}{g^3}\right)\frac{dy}{dx}\right.$$
$$\left. + \left(-\frac{8}{3}\frac{g''^2}{g^2} + \frac{g''''}{g} - \frac{14}{3}\frac{g'g'''}{g^2} + \frac{38}{3}\frac{g'^2 g''}{g^3} - \frac{56}{9}\frac{g'^4}{g^4}\right)y\right\}. \tag{3.37}$$

From the equation

$$\frac{d^4 y_1}{dx_1^4} + \frac{d}{dx_1}\left(p(x_1)\frac{dy_1}{dx_1}\right) + q(x_1)y_1 = 0,$$

and (3.37) we deduce that if the resulting equation is to be identical with (3.34) then we require $f(x)$ and $g(x)$ to satisfy

$$p(x) = p(f)g^{4/3} + \frac{10}{3}\left(\frac{g''}{g} - \frac{4}{3}\frac{g'^2}{g^2}\right),$$

$$q(x) = q(f)g^{8/3} + p(f)\left(g^{1/3}g'' - \frac{2}{3}\frac{g'^2}{g^{2/3}}\right) + \frac{dp}{df}gg' \qquad (3.38)$$

$$+ \left(\frac{g''''}{g} - \frac{14}{3}\frac{g'g'''}{g^2} - \frac{8}{3}\frac{g''^2}{g^2} + \frac{38}{3}\frac{g'^2 g''}{g^3} - \frac{56}{9}\frac{g'^4}{g^4}\right)$$

From (3.6), (3.36) and (3.38) we have,

$$5\xi''' + 2p\xi' + p'\xi = 0,$$

$$(3.39)$$

$$3[\xi'''' + p\xi'']' + 8q\xi' + 2q'\xi = 0,$$

and $\eta = 3\xi'/2$. The two equations (3.39) are consistent only if

$$\xi^4\left(q - \left(\frac{3p}{10}\right)^2 - \frac{3p''}{10}\right) = D, \qquad (3.40)$$

where D is a constant. Thus in general for a given $p(x)$ we need to assume $q(x)$ is given by

$$q(x) = \frac{9p(x)^2}{100} + \frac{3}{10}\frac{d^2 p}{dx^2} + \frac{D}{\xi(x)^4}, \qquad (3.41)$$

where $\xi(x)$ is a solution of $(3.39)_1$ and has the general form (3.30) where $\phi_1(x)$ and $\phi_2(x)$ are linearly independent solutions of (3.35). Moreover if the constants A, B and C are as in (3.30) then we have

$$\frac{1}{4}(2\xi\xi'' - \xi'^2) + \frac{p}{10}\xi^2 = (AC - B^2). \qquad (3.42)$$

The global form of the one-parameter group (3.1) can be deduced from

$$\frac{dx_1}{d\epsilon} = \xi(x_1), \quad \frac{dy_1}{d\epsilon} = \frac{3}{2}\xi'(x_1)y_1,$$

subject to the initial conditions $x_1 = x$, $y_1 = y$ when $\epsilon = 0$. Suitable canonical coordinates (u, v) are given by

$$u(x,y) = \frac{y}{\xi(x)^{3/2}}, \quad v(x,y) = \int_{x_0}^{x} \frac{dt}{\xi(t)},$$

where x_0 is some constant. Making use of the result given in Problem 18 (replacing $\alpha(x)$ and $\beta'(x)$ with $\xi(x)^{3/2}$ and $\xi(x)^{-1}$ respectively) we find that in terms of (u, v) the differential equation (3.34) eventually becomes

$$\frac{d^4u}{dv^4} + 10(AC - B^2)\frac{d^2u}{dv^2} + [D + 9(AC - B^2)^2]u = 0, \qquad (3.43)$$

where we have made use of $(3.39)_1$, (3.40) and (3.42). Thus provided $q(x)$ is given by (3.41) then (3.34) can be reduced to a linear equation with constant coefficients. Evidently the approach presupposes that the linearly independent solutions $\phi_1(x)$ and $\phi_2(x)$ of the associated equation (3.35) can be readily obtained.

Example 3.3 Illustrate the above procedure for the case of $p(x)$ identically zero.

In this case the linearly independent solutions of (3.35) are $\phi_1(x) = 1$ and $\phi_2(x) = x$. Hence $\xi(x)$ is of the form,

$$\xi(x) = A + 2Bx + Cx^2,$$

and the above approach is effective provided that $q(x)$ is given by

$$q(x) = \frac{D}{(A + 2Bx + Cx^2)^4},$$

from some constants A, B, C and D. For purposes of illustration suppose that $A = 0$, $B = -\frac{1}{2}$ and $C = D = 1$. In this case we have

$$\xi(x) = x(x - 1), \quad q(x) = [x(1 - x)]^{-4},$$

and (3.43) has the general solution

$$u = (C_1 + C_2 v)e^{v\sqrt{5}/2} + (C_3 + C_4 v)e^{-v\sqrt{5}/2},$$

where C_1, C_2, C_3 and C_4 denote four arbitrary constants. The solution of the original equation of the form (3.34) can now be readily deduced.

[In the following problems $s(x)$ is assumed defined by (3.4). Also for Problems 4, 5, 6 and 7 a further arbitrary constant could be introduced into the condition restricting the coefficients. In these problems we have assumed that this constant has been absorbed into the constant x_0 in (3.4) which defines $s(x)$.]

Invariance of Standard Linear Ordinary Differential Equations

PROBLEMS

1. For Bernoulli's equation,

$$\frac{dy}{dx} + p(x)y = q(x)y^n \quad (n \neq 1),$$

show that

$$\xi(x) = \frac{1}{q(x)s(x)^{1-n}} \left\{ (1-n)C_1 \int_{x_0}^{x} s(t)^{1-n} q(t) dt + C_2 \right\},$$

$$\eta(x) = C_1 - p(x)\xi(x).$$

2. **Continuation.** If the constant C_1 is non-zero deduce that suitable canonical coordinates (u, v) are

$$u(x, y) = \frac{s(x)y}{\left\{ (1-n)C_1 \int_{x_0}^{x} s(t)^{1-n} q(t) dt + C_2 \right\}^{\frac{1}{(1-n)}}},$$

$$v(x, y) = \frac{1}{(1-n)C_1} \log \left\{ (1-n)C_1 \int_{x_0}^{x} s(t)^{1-n} q(t) dt + C_2 \right\},$$

and therefore the differential equation becomes

$$\frac{du}{dv} = u(u^{n-1} - C_1).$$

Integrate this as a separable equation to obtain

$$(1 - C_1 u^{1-n}) e^{(1-n)C_1 v} = C_3,$$

and hence deduce the solution of the original equation.

3. **Continuation.** If the constant C_1 is zero show that

$$u(x, y) = s(x)y, \quad v(x, y) = \frac{1}{C_2} \int_{x_0}^{x} s(t)^{1-n} q(t) dt,$$

and that the differential equation becomes

$$\frac{du}{dv} = C_2 u^n.$$

Integrate to obtain

$$u^{1-n} - C_2(1-n)v = C_4,$$

and show that the same solution is obtained as in the previous problem with $C_4 = (C_2 - C_3)/C_1$.

4. Show that the generalised Riccati equation

$$\frac{dy}{dx} + p(x)y = q(x) + r(x)y^2,$$

remains invariant under (3.1) provided $r(x) = q(x)s(x)^2$. If this is the case, show that

$$\xi(x) = \frac{1}{q(x)s(x)}, \quad \eta(x) = \frac{-p(x)}{q(x)s(x)},$$

and that suitable canonical coordinates are,

$$u(x,y) = s(x)y, \quad v(x,y) = \int_{x_0}^{x} s(t)q(t)dt.$$

Hence show that the differential equation becomes

$$\frac{du}{dv} = 1 + u^2,$$

and therefore the solution of the original equation is

$$y(x) = \frac{1}{s(x)} \tan\left(\int_{x_0}^{x} s(t)q(t)dt + C\right).$$

5. Show that the Abel equation of the first kind,

$$\frac{dy}{dx} + p(x)y = q(x) + r(x)y^3,$$

is invariant under the group of the previous problem provided $r(x) = q(x)s(x)^3$. Show that the differential equation becomes

$$\frac{du}{dv} = 1 + u^3.$$

6. Show that

$$\frac{dy}{dx} + p(x)y = q(x) + r(x)\log y,$$

is invariant under (3.1) provided $q(x) = r(x)\log s(x)$. Show that

$$\xi(x) = \frac{1}{r(x)s(x)}, \quad \eta(x) = \frac{-p(x)}{r(x)s(x)},$$

$$u(x,y) = s(x)y, \quad v(x,y) = \int_{x_0}^{x} s(t)r(t)dt,$$

and that the differential equation becomes

$$\frac{du}{dv} = \log u.$$

7. Verify that the differential equation

$$\frac{dy}{dx} + p(x)y = q(x)y^m + r(x)y^n,$$

admits the group (3.1) provided $r(x) = q(x)s(x)^{n-m}$. If this is the case deduce that,

$$\xi(x) = \frac{1}{q(x)s(x)^{1-m}}, \quad \eta(x) = \frac{-p(x)}{q(x)s(x)^{1-m}},$$

$$u(x,y) = s(x)y, \quad v(x,y) = \int_{x_0}^{x} s(t)^{1-m} q(t)dt,$$

and that the differential equation becomes

$$\frac{du}{dv} = u^m + u^n.$$

8. Show that the linear homogeneous second order equation

$$\frac{d^2y}{dx^2} + a(x)\frac{dy}{dx} + b(x)y = 0,$$

can be reduced to normal form either by,
(i) changing the dependent variable to y^* where

$$y = e^{-\frac{1}{2}\int_{x_0}^{x} a(t)dt} y^*,$$

in which case we have

$$\frac{d^2y^*}{dx^2} + \left\{ b(x) - \frac{a(x)^2}{4} - \frac{1}{2}\frac{da}{dx} \right\} y^* = 0,$$

(ii) changing the independent variable to x^* where

$$x^* = \int_{x_0}^{x} e^{-\int_{x_0}^{s} a(t)dt} ds,$$

in which case we have

$$\frac{d^2y}{dx^{*2}} + \frac{b(x)}{\left(\frac{dx^*}{dx}\right)^2} y = 0.$$

9. **Continuation.** For the equation,

$$\frac{d^2y}{dx^2} - \frac{3x}{(1-x^2)}\frac{dy}{dx} + \frac{n(n+2)}{(1-x^2)}y = 0,$$

show that the reductions to normal form given in the previous problem give rise to the following equations,

(i) $$\frac{d^2y^*}{dx^2} + \left\{\frac{n(n+2)}{(1-x^2)} + \frac{3}{4}\frac{(2-x^2)}{(1-x^2)^2}\right\} y^* = 0,$$

(ii) $$\frac{d^2y}{dx^{*2}} + \frac{n(n+2)}{(1+x^{*2})^2} y = 0.$$

10. With the notation of Section 3.3 consider the non-homogeneous equation

$$\frac{d^2y}{dx^2} + p(x)y = q(x).$$

If this equation is to remain invariant under the same group which leaves the homogeneous equation unaltered then show that $q(x)$ must be given by

$$q(x) = q_0 \xi(x)^{-3/2},$$

where q_0 is a constant. Hence show that the equation corresponding to (3.21) becomes

$$\frac{d^2u}{dv^2} + (AC - B^2)u = q_0.$$

11. If $\xi(x)$ satisfies

$$\frac{1}{4}(2\xi\xi'' - \xi'^2) + p\xi^2 = K^2,$$

where K is a constant, show that $w(x) = \xi(x)^{1/2}$ satisfies the non-linear second order equation

$$\frac{d^2w}{dx^2} + p(x)w = \frac{K^2}{w^3}.$$

Invariance of Standard Linear Ordinary Differential Equations

12. The following three operations leave a linear differential equation linear,
    (i) changing the dependent variable to y^* where $y = \alpha(x)y^*$,
    (ii) changing the independent variable to x^* where $x^* = \beta(x)$,
    (iii) multiplication of the equation by a non-zero function $\gamma(x)$.

    Show that by choosing α, β and γ such that

    $$\alpha(x)\beta'(x) = e^{-\int_{x_0}^{x} \frac{B(t)}{3A(t)} dt}, \quad \gamma(x) = [\alpha(x)\beta'(x)^3 A(x)]^{-1},$$

    the general linear third order equation

    $$A(x)\frac{d^3y}{dx^3} + B(x)\frac{d^2y}{dx^2} + C(x)\frac{dy}{dx} + D(x)y = 0, \qquad (*)$$

    can be reduced to an equation of the form,

    $$\frac{d^3y}{dx^3} + a(x)\frac{dy}{dx} + b(x)y = 0. \qquad (**)$$

13. **Continuation.** A second order linear equation is self-adjoint if it is of the form

    $$\frac{d}{dx}\left(P(x)\frac{dy}{dx}\right) + Q(x)y = 0.$$

    Show that any second-order linear differential equation can be made self-adjoint by any one of the operations (i), (ii) and (iii) of the previous problem.

14. **Continuation.** A third order equation is formally self-adjoint (or anti self-adjoint) if it has the form (Murphy (1960), page 199)

    $$\frac{d^2}{dx^2}\left(P(x)\frac{dy}{dx}\right) + \frac{d}{dx}\left(P(x)\frac{d^2y}{dx^2}\right) + \frac{d}{dx}(Q(x)y) + Q(x)\frac{dy}{dx} = 0.$$

    Show that the general equation (*) is self-adjoint if and only if

    $$B(x) = \frac{3}{2}\frac{dA}{dx}, \quad D(x) = \frac{1}{2}\frac{d}{dx}\left\{C(x) - \frac{1}{3}\frac{dB}{dx}\right\}.$$

    Make use of this result and the reduced equation (**) to show that no combination of the operations (i), (ii) and (iii) of Problem 12 can make a third order equation self-adjoint unless it is originally self-adjoint.

15. **Continuation.** Show that if a third order equation is self-adjoint then it remains self-adjoint under (i), (ii) and (iii) of Problem 12 provided $\alpha(x)$ is a constant multiple of $\gamma(x)\beta'(x)$.

16. **Continuation.** Show using the operations of Problem 12, that the general fourth order equation

$$A(x)\frac{d^4y}{dx^4} + B(x)\frac{d^3y}{dx^3} + C(x)\frac{d^2y}{dx^2} + D(x)\frac{dy}{dx} + E(x)y = 0, \qquad (+)$$

can be reduced to one of the form,

$$\frac{d^4y}{dx^4} + a(x)\frac{d^2y}{dx^2} + b(x)\frac{dy}{dx} + c(x)y = 0, \qquad (++)$$

by choosing α, β and γ to be such that

$$\alpha(x)^2 \beta'(x)^3 = e^{-\int_{x_0}^{x} \frac{B(t)}{2A(t)} dt}, \qquad \gamma(x) = [\alpha(x)\beta'(x)^4 A(x)]^{-1}.$$

17. **Continuation.** A fourth order equation is self-adjoint if it has the form

$$\frac{d^2}{dx^2}\left(P(x)\frac{d^2y}{dx^2}\right) + \frac{d}{dx}\left(Q(x)\frac{dy}{dx}\right) + R(x)y = 0.$$

Show that $(+)$ is self-adjoint if and only if

$$B(x) = 2\frac{dA}{dx}, \qquad D(x) = \frac{d}{dx}\left\{C(x) - \frac{1}{2}\frac{dB}{dx}\right\}.$$

18. **Continuation.** Show that no combination of (i), (ii) and (iii) of Problem 12 can make $(++)$ self-adjoint unless it is self-adjoint originally. If $(++)$ is self-adjoint show that these operations give rise to another self-adjoint equation provided $\alpha(x)$ is a constant multiple of $\gamma(x)\beta'(x)$.

[Hint, for the second part, if $(++)$ is self-adjoint we have $b(x) = a'(x)$ and the equation becomes

$$\frac{d^2}{dx^{*2}}\left(P^* \frac{d^2y^*}{dx^{*2}}\right) + \frac{d}{dx^*}\left(Q^* \frac{dy^*}{dx^*}\right) + R^* y^* = 0, \qquad (+++)$$

where,

$$P^* = \alpha^2 \beta'^3,$$

$$Q^* = \alpha^2 \beta''' + 2\alpha\alpha' \beta'' + (4\alpha\alpha'' - 2\alpha'^2 + a\alpha^2)\beta',$$

$$R^* = \frac{\alpha}{\beta'}(\alpha'''' + a\alpha'' + a'\alpha' + c\alpha),$$

where α, β, a and c are all functions of x, primes denote differentiation with respect to x and we have taken $\gamma = \alpha/\beta'$.]

19. **Continuation.** If in the previous problem the functions $a(x)$ and $c(x)$ are such that

$$c = \frac{a''}{2} + \left(\frac{a}{2}\right)^2,$$

show that $(+++)$ admits the factorization

$$L^2[\lambda_3 L^2 y^*] = 0,$$

where L^2 is the second order operator defined by

$$L^2 y^* = \frac{d}{dx^*}\left(\lambda_1 \frac{dy^*}{dx^*}\right) + \lambda_2 y^*,$$

where λ_1, λ_2 and λ_3 are given by

$$\lambda_1 = \alpha^2 \beta', \quad \lambda_2 = \frac{\alpha}{\beta'}\left(\alpha'' + \frac{a}{2}\alpha\right), \quad \lambda_3 = \frac{\beta'}{\alpha^2}.$$

20. Verify by differentiation that the third order self-adjoint equation of Problem 14 admits the first integral

$$P(2yy'' - y'^2) + P'yy' + Qy^2 = \text{constant}.$$

If $\phi_1(x)$ and $\phi_2(x)$ are linearly independent solutions of the second order equation

$$4P\frac{d^2 y}{dx^2} + 2P'\frac{dy}{dx} + Qy = 0,$$

and if $y(x)$ is given by

$$y = A\phi_1^2 + 2B\phi_1\phi_2 + C\phi_2^2,$$

where A, B and C denote arbitrary constants then deduce that

$$P(2yy'' - y'^2) + P'yy' + Qy^2 = 4(AC - B^2)P\omega^2,$$

where $\omega(x)$ is the Wronskian of ϕ_1 and ϕ_2, namely

$$\omega = \phi_1 \phi_2' - \phi_2 \phi_1'.$$

Hence conclude that this expression for y gives the general solution to the self-adjoint equation of Problem 14.

21. In the notation of Section 3.4:

   (i) Deduce from (3.26) and (3.27) the equation,
   $$\left(q(f) - \frac{1}{2}\frac{dp(f)}{df}\right)g(x)^3 = \left(q(x) - \frac{1}{2}\frac{dp(x)}{dx}\right),$$
   and hence deduce the condition (3.29).

   (ii) Consider the non-homogeneous equation
   $$\frac{d^3y}{dx^3} + p(x)\frac{dy}{dx} + q(x)y = r(x),$$
   and show that this equation remains invariant under the same group which leaves the homogeneous equation unaltered provided
   $$r(x) = r_0\xi(x)^{-2},$$
   where r_0 is a constant. Hence show that the equation corresponding to (3.33) becomes
   $$\frac{d^3u}{dv^3} + 4(AC - B^2)\frac{du}{dv} + Du = r_0.$$

22. With the notation of Section 3.5:

   (i) Deduce from (3.36) and (3.38) the equation,
   $$\left(q(f) - \frac{9p(f)^2}{100} - \frac{3}{10}\frac{d^2p(f)}{df^2}\right)g(x)^{8/3} = \left(q(x) - \frac{9p(x)^2}{100} - \frac{3}{10}\frac{d^2p(x)}{dx^2}\right),$$
   and hence deduce the condition (3.40).

   (ii) Consider the non-homogeneous equation
   $$\frac{d^4y}{dx^4} + \frac{d}{dx}\left(p(x)\frac{dy}{dx}\right) + q(x)y = r(x),$$
   and show that this equation remains invariant under the same group which leaves the homogeneous equation unaltered provided
   $$r(x) = r_0\xi(x)^{-5/2},$$
   where r_0 is a constant. Hence show that the equation corresponding to (3.43) becomes
   $$\frac{d^4u}{dv^4} + 10(AC - B^2)\frac{d^2u}{dv^2} + [D + 9(AC - B^2)^2]u = r_0.$$

Chapter Four
First order ordinary differential equations

4.1 Introduction

In this chapter we discuss *Lie's fundamental problem* of finding a one-parameter group which leaves a given first order ordinary differential equation unaltered. That is, for a given $F(x,y)$ we wish to determine a one-parameter group,

$$x_1 = x + \epsilon\xi(x,y) + \mathbf{O}(\epsilon^2), \quad y_1 = y + \epsilon\eta(x,y) + \mathbf{O}(\epsilon^2), \tag{4.1}$$

such that the differential equation,

$$\frac{dy}{dx} = F(x,y), \tag{4.2}$$

remains invariant. This problem is by no means solved. Much of the literature is concerned with the *alternative problem* of finding differential equations which are left invariant by a given one-parameter group. For this aspect the reader should consult the standard tables of differential equations and their associated groups (see for example either Dickson (1924), page 324 or Bluman and Cole (1974), page 99). We shall also consider the alternative problem but with a view to situations not previously discussed. For the fundamental problem we highlight the role of singular and special solutions of (4.2) and we refer the reader to the related discussion given by Page (1897)(page 113).

Integral curves of (4.2) $z(x,y) =$ constant, evidently satisfy the first order partial differential equation

$$\frac{\partial z}{\partial x} + F(x,y)\frac{\partial z}{\partial y} = 0. \tag{4.3}$$

In Section 4.6 we consider the invariance of (4.3) under a one-parameter group in the three variables (x,y,z) which we relate to integrating factors of (4.2). In a sense this result provides a generalization of Lie's famous result for integrating factors (see Problem 1). This section deals briefly with the group approach to partial differential equations and therefore the reader is perhaps best advised to avoid it until familiar with the material on partial differential equations described in subsequent chapters. In the final section of this chapter we attempt the solution of Lie's fundamental problem. Since the two functions $\xi(x,y)$ and $\eta(x,y)$ are not completely determined by the single constraint (4.6) Lie's problem is rather to propose a second independent constraint on the group (4.1) which is in some sense compatible with (4.6) so as to simplify the subsequent analysis. Here we propose that the assumption that (4.1) is area preserving may be such a constraint.

Although the results obtained are by no means conclusive, different forms of Lie's problem are generated which at least convey some insight into the fundamental difficulties associated with the problem.

4.2 Infinitesimal versions of y' and $y' = F(x, y)$ and the fundamental problem

We calculate the infinitesimal version of y' as follows. From (4.1) we have,

$$\frac{dy_1}{dx_1} = \frac{dy + \epsilon\left(\frac{\partial \eta}{\partial x}dx + \frac{\partial \eta}{\partial y}dy\right)}{dx + \epsilon\left(\frac{\partial \xi}{\partial x}dx + \frac{\partial \xi}{\partial y}dy\right)} + \mathbf{O}(\epsilon^2),$$

and on dividing through by dx we obtain,

$$\frac{dy_1}{dx_1} = \frac{\frac{dy}{dx} + \epsilon\left(\frac{\partial \eta}{\partial x} + \frac{\partial \eta}{\partial y}\frac{dy}{dx}\right)}{1 + \epsilon\left(\frac{\partial \xi}{\partial x} + \frac{\partial \xi}{\partial y}\frac{dy}{dx}\right)} + \mathbf{O}(\epsilon^2).$$

Hence on using the binomial theorem for the denominator we have

$$\frac{dy_1}{dx_1} = \frac{dy}{dx} + \epsilon \pi(x, y, y') + \mathbf{O}(\epsilon^2), \tag{4.4}$$

where $\pi(x, y, y')$ is given by

$$\pi = \frac{\partial \eta}{\partial x} + \left(\frac{\partial \eta}{\partial y} - \frac{\partial \xi}{\partial x}\right)\frac{dy}{dx} - \frac{\partial \xi}{\partial y}\left(\frac{dy}{dx}\right)^2, \tag{4.5}$$

and this is the infinitesimal version of y'.

If (4.1) leaves (4.2) invariant then from (4.4),

$$\frac{dy_1}{dx_1} = F(x_1, y_1),$$

and

$$F(x_1, y_1) = F(x, y) + \epsilon\left(\xi\frac{\partial F}{\partial x} + \eta\frac{\partial F}{\partial y}\right) + \mathbf{O}(\epsilon^2),$$

we obtain

$$\frac{dy}{dx} + \epsilon \pi(x, y, y') = F(x, y) + \epsilon\left(\xi\frac{\partial F}{\partial x} + \eta\frac{\partial F}{\partial y}\right) + \mathbf{O}(\epsilon^2),$$

First Order Ordinary Differential Equations 53

and therefore from the terms of order ϵ we have

$$\xi \frac{\partial F}{\partial x} + \eta \frac{\partial F}{\partial y} = \frac{\partial \eta}{\partial x} + \left(\frac{\partial \eta}{\partial y} - \frac{\partial \xi}{\partial x}\right) F - \frac{\partial \xi}{\partial y} F^2, \tag{4.6}$$

where we have used (4.1) and (4.5). *Lie's fundamental problem for first order differential equations is that for a given $F(x,y)$ how can we systematically determine two functions $\xi(x,y)$ and $\eta(x,y)$ such that (4.6) is satisfied.* The functions $\xi(x,y)$ and $\eta(x,y)$ can be completely arbitrary provided (4.6) is satisfied and $\eta \neq F\xi$. Equation (4.6) always admits the solution $\eta = F\xi$. However, this solution does not serve our purposes since in this case when we come to deduce the global form of (4.1) we need to solve

$$\frac{dx_1}{d\epsilon} = \xi(x_1, y_1), \quad \frac{dy_1}{d\epsilon} = F(x_1, y_1)\xi(x_1, y_1),$$

and thus we are led back to our original problem (4.2). Further from (4.8) in the following section, it is also evident that $\eta = F\xi$ is not an acceptable solution of (4.6).

If $\xi(x,y)$ and $\eta(x,y)$ are known functions then we show in the following section that the condition (4.6) reduces to the existence of an integrating factor for the differential equation (4.2). Moreover for given $\xi(x,y)$ and $\eta(x,y)$ we may view (4.6) as a first order partial differential equation for the determination of $F(x,y)$. Thus we may determine classes of differential equations invariant under a known one-parameter group and this is the *alternative problem* which is discussed in the section thereafter.

4.3 Integrating factors and canonical coordinates for $y' = F(x,y)$

If we introduce $\lambda(x,y)$ by $\lambda = \eta - F\xi$ then (4.6) simplifies considerably and we obtain

$$\frac{\partial \lambda}{\partial x} + F^2 \frac{\partial}{\partial y}\left(\frac{\lambda}{F}\right) = 0. \tag{4.7}$$

Although this equation is a good deal simpler than (4.6), the interesting aspect of (4.6) has been removed since (4.7) does not involve either ξ or η directly. If we introduce $\mu(x,y)$ by $\mu = \lambda^{-1}$ then we have

$$\mu(x,y) = \frac{1}{\eta(x,y) - F(x,y)\xi(x,y)}, \tag{4.8}$$

and (4.7) can be shown to become,

$$\frac{\partial \mu}{\partial x} + \frac{\partial}{\partial y}(F\mu) = 0. \tag{4.9}$$

Hence if we write the original differential equation (4.2) as

$$dy - F(x,y)dx = 0, \qquad (4.10)$$

then from (4.9) we see that $\mu(x,y)$ is an integrating factor for (4.10). This result is due originally to Lie (see Problem 1) and is generally given some prominence in the literature. However, from the point of view of actually solving differential equations the use of canonical coordinates is preferable. Moreover as we have seen in the previous chapter canonical coordinates can be used with higher order equations and therefore we will emphasise their use here.

From (4.8) and (4.9) we see that there exists a function $z(x,y)$ such that

$$\frac{\partial z}{\partial x} = \frac{-F}{(\eta - F\xi)}, \quad \frac{\partial z}{\partial y} = \frac{1}{(\eta - F\xi)}. \qquad (4.11)$$

But we have

$$dz = \frac{\partial z}{\partial x}dx + \frac{\partial z}{\partial y}dy = \frac{dy - Fdx}{(\eta - F\xi)} = 0,$$

where we have used (4.10) and (4.11). Thus $z(x,y) = C$ where C is a constant represents the integral of (4.2) and we have using (4.11)

$$\xi\frac{\partial z}{\partial x} + \eta\frac{\partial z}{\partial y} = 1. \qquad (4.12)$$

Thus if we introduce the operator L by

$$L = \xi\frac{\partial}{\partial x} + \eta\frac{\partial}{\partial y},$$

then (4.12) gives $L(z) = 1$ and from the Commutation theorem (see (2.25)) we have

$$z(x_1, y_1) = e^{\epsilon L} z(x,y) = \sum_{n=0}^{\infty} \frac{\epsilon^n}{n!} L^n(z),$$

and therefore

$$z(x_1, y_1) = z(x,y) + \epsilon. \qquad (4.13)$$

From this equation and (2.7) we see that if a first order differential equation is invariant under a one-parameter group then the required integral has the form

$$z(x,y) = v(x,y) + \psi[u(x,y)], \qquad (4.14)$$

where (u,v) are the canonical coordinates of the group and ψ is some function of u only.

First Order Ordinary Differential Equations

In order to obtain (4.14) more directly we suppose that in terms of canonical coordinates the differential equation (4.2) becomes

$$\frac{dv}{du} = \phi(u,v).$$

But clearly if this equation is invariant under $u_1 = u$ and $v_1 = v + \epsilon$ then ϕ must be independent of v and the result (4.14) follows immediately from the equation

$$\frac{dv}{du} = \phi(u).$$

Example 4.1 Solve the differential equation,

$$\frac{dy}{dx} = \frac{y}{(x + x^2 + y^2)},$$

by finding a one-parameter group leaving it invariant.

In this case we have

$$\frac{\partial F}{\partial x} = \frac{-(1+2x)y}{(x+x^2+y^2)^2}, \quad \frac{\partial F}{\partial y} = \frac{(x+x^2-y^2)}{(x+x^2+y^2)^2},$$

and from (4.6) we need to find $\xi(x,y)$ and $\eta(x,y)$ such that

$$(x+x^2-y^2)\eta - (1+2x)y\xi = (x+x^2+y^2)^2 \frac{\partial \eta}{\partial x} + \left(\frac{\partial \eta}{\partial y} - \frac{\partial \xi}{\partial x}\right) y(x+x^2+y^2) - y^2 \frac{\partial \xi}{\partial y}.$$

Unfortunately $\xi(x,y)$ and $\eta(x,y)$ must now be determined by trial and error. Try $\eta = 1$ (that is, η constant) then ξ must satisfy

$$(x+x^2+y^2)\frac{\partial \xi}{\partial x} + y\frac{\partial \xi}{\partial y} = (1+2x)\xi - \frac{(x+x^2-y^2)}{y}.$$

Unfortunately even at this stage we cannot systematically solve this equation since the solution by Lagrange's method involves solving the original differential equation (see Problem 3). However, with some persistence we can arrive at the solution $\xi = x/y$. Thus the global form of the one-parameter group is obtained by solving

$$\frac{dx_1}{d\epsilon} = \frac{x_1}{y_1}, \quad \frac{dy_1}{d\epsilon} = 1,$$

subject to the initial conditions $x_1 = x$, $y_1 = y$ when $\epsilon = 0$. We obtain

$$x_1 = x + \epsilon \frac{x}{y}, \quad y_1 = y + \epsilon,$$

and the reader should verify that the given differential equation is indeed invariant under this group. Canonical coordinates (u,v) are given by

$$u = \frac{y}{x}, \quad v = y,$$

and the differential equation becomes

$$\frac{dv}{du} = \frac{1}{\left(\frac{1}{x} - \frac{y}{x^2}\frac{dx}{dy}\right)} = -\frac{1}{(1+u^2)}.$$

Thus the solution is

$$v + \tan^{-1} u = C,$$

or

$$y + \tan^{-1} \frac{y}{x} = C.$$

Alternatively if we write the differential equation as (4.10), namely

$$dy - \frac{ydx}{(x+x^2+y^2)} = 0,$$

then (4.8) gives the integrating factor $\mu(x,y)$ as

$$\mu(x,y) = \frac{(x+x^2+y^2)}{(x^2+y^2)},$$

and we obtain

$$dy + \frac{(xdy - ydx)}{(x^2+y^2)} = 0.$$

This integrates to give the previously obtained result.

Example 4.2 Obtain a function $a(x)$ or class of functions such that the differential equation

$$\frac{dy}{dx} = a(x) + y^2,$$

remains invariant under a one-parameter group.

First Order Ordinary Differential Equations

From (4.6) we have

$$\xi a' + 2y\eta = \frac{\partial \eta}{\partial x} + \left(\frac{\partial \eta}{\partial y} - \frac{\partial \xi}{\partial x}\right)(a + y^2) - \frac{\partial \xi}{\partial y}(a^2 + 2ay^2 + y^4).$$

This condition simplifies if $\xi = \xi(x)$ and $\eta = \eta(x)y$ and we obtain

$$\xi(x)a'(x) + 2\eta(x)y^2 = \eta'(x)y + [\eta(x) - \xi'(x)][a(x) + y^2].$$

From this equation we deduce on equating coefficients of powers of y,

$$\xi(x) = Ax + B, \quad \eta(x) = -A,$$

provided $a(x)$ takes the form,

$$a(x) = \frac{C}{(Ax+B)^2},$$

where A, B and C are all constants. From

$$\frac{dx_1}{d\epsilon} = (Ax_1 + B), \quad \frac{dy_1}{d\epsilon} = -Ay_1,$$

we deduce that suitable canonical coordinates (u, v) are given by

$$u = (Ax + B)y, \quad v = \frac{1}{A}\log(Ax + B),$$

and from the original differential equation we obtain

$$\frac{dv}{du} = \frac{1}{(u^2 + Au + C)}.$$

This equation is separable and can be readily integrated for given values of the constants A and C.

4.4 The alternative problem

For a given $\xi(x, y)$ and $\eta(x, y)$ can we obtain the most general $F(x, y)$ such that (4.6) is satisfied? We solve (4.6) as a first order partial differential equation in F (see Problem 3). The characteristic equations are

$$\frac{dx}{d\tau} = \xi(x, y), \quad \frac{dy}{d\tau} = \eta(x, y), \tag{4.15}$$

and

$$\frac{dF}{d\tau} = \frac{\partial \eta}{\partial x} + \left(\frac{\partial \eta}{\partial y} - \frac{\partial \xi}{\partial x}\right)F - \frac{\partial \xi}{\partial y}F^2, \tag{4.16}$$

and in order to obtain the most general $F(x,y)$ we need to deduce two independent integrals of (4.15) and (4.16). In general (4.16) is a Riccati equation which we solve using the known solution of (4.6), namely $\eta = F\xi$. Making the substitution (see Problem 4)

$$F = \frac{\eta}{\xi} + \frac{1}{w}, \tag{4.17}$$

we obtain

$$\frac{dw}{d\tau} + \left(\frac{\partial \eta}{\partial y} - \frac{\partial \xi}{\partial x} - 2\frac{\eta}{\xi}\frac{\partial \xi}{\partial y}\right)w = \frac{\partial \xi}{\partial y},$$

which is linear and can be solved in the usual way.

Example 4.3 Obtain the most general first order differential equation invariant under a one-parameter group of the form,

$$x_1 = f(x), \quad y_1 = g(x)y.$$

Infinitesimally we have

$$x_1 = x + \epsilon\xi(x) + \mathbf{O}(\epsilon^2), \quad y_1 = y + \epsilon\eta(x)y + \mathbf{O}(\epsilon^2),$$

and therefore the characteristic equations (4.15) and (4.16) become

$$\frac{dx}{d\tau} = \xi(x), \quad \frac{dy}{d\tau} = \eta(x)y, \tag{4.19}$$

and

$$\frac{dF}{d\tau} = \eta'(x)y + (\eta(x) - \xi'(x))F. \tag{4.20}$$

From (4.19) we have

$$\frac{dy}{dx} = \frac{\eta(x)}{\xi(x)}y,$$

and therefore

$$ys(x) = A, \tag{4.21}$$

First Order Ordinary Differential Equations

where A is a constant and $s(x)$ is defined by

$$s(x) = e^{-\int_{x_0}^{x} \frac{\eta(t)}{\xi(t)} dt}, \qquad (4.22)$$

for some constant x_0. From (4.19)$_1$, (4.20) and (4.21) we obtain

$$\frac{dF}{dx} + \left(\frac{\xi'(x)}{\xi(x)} - \frac{\eta(x)}{\xi(x)}\right) F = \frac{A\eta'(x)}{\xi(x)s(x)},$$

which integrates to give

$$\xi(x)s(x)F = A\eta(x) + B, \qquad (4.23)$$

where B is a constant. Hence our most general first order differential equation is obtained from $B = \Phi(A)$, that is

$$\frac{dy}{dx} - \frac{\eta(x)}{\xi(x)} y = \frac{\Phi[s(x)y]}{\xi(x)s(x)}.$$

In this case we can verify that suitable canonical coordinates (u, v) are given by

$$u(x, y) = s(x)y, \quad v(x, y) = \int_{x_0}^{x} \frac{dt}{\xi(t)},$$

and that the differential equation becomes

$$\frac{du}{dv} = \Phi(u).$$

[Notice this example generalizes Problems 4, 5, 6 and 7 of Chapter 3.]

Example 4.4 Obtain the most general first order differential equation invariant under the one-parameter group,

$$\xi(x, y) = \xi(x)e^{ky}, \quad \eta(x, y) = \eta(x)e^{ky},$$

where k is a constant.

In this case we have from (4.15)

$$\frac{dy}{dx} = \frac{\eta(x)}{\xi(x)},$$

and therefore

$$y - \int_{x_0}^{x} \frac{\eta(t)}{\xi(t)} dt = A, \qquad (4.24)$$

where A is a constant. Making use of the result given in Problem 6 we have on performing the integration

$$W = \frac{k}{e^{kA}} \int_{x_0}^{x} \frac{s(t)^k}{\xi(t)} dt + B, \qquad (4.25)$$

where B is a constant and $s(x)$ is defined by (4.22). Since $w = \xi(x,y)W$ we have from (4.17)

$$W = \frac{1}{e^{ky}} \left(\xi(x) \frac{dy}{dx} - \eta(x) \right)^{-1},$$

and hence the required differential equation is

$$\left(\xi(x) \frac{dy}{dx} - \eta(x) \right)^{-1} = \frac{k}{s(x)^k} \int_{x_0}^{x} \frac{s(t)^k}{\xi(t)} dt + \Phi \left(y - \int_{x_0}^{x} \frac{\eta(t)}{\xi(t)} dt \right),$$

where Φ denotes an arbitrary function. Suitable canonical coordinates are

$$u(x,y) = y - \int_{x_0}^{x} \frac{\eta(t)}{\xi(t)} dt, \quad v(x,y) = \frac{1}{s(x)^k e^{ky}} \int_{x_0}^{x} \frac{s(t)^k}{\xi(t)} dt,$$

and on using

$$\frac{du}{dv} = \frac{du/dx}{dv/dx} = \frac{1}{\left(-kv + \frac{1}{e^{ky}\xi(x)} dx/du \right)},$$

we see that the differential equation becomes

$$\frac{du}{dv} = \frac{e^{ku}}{\Phi(u)}.$$

The remaining sections of this chapter are devoted to various aspects associated with Lie's fundamental problem and the condition (4.6).

4.5 The fundamental problem and singular solutions of $y' = F(x,y)$

Suppose the integral $z(x,y) = C$ of (4.2) is solvable for y so that we have

$$y = S(x, C). \qquad (4.26)$$

But from (4.13) we see that if (4.2) is invariant under the one-parameter group (4.1) then

$$y_1 = S(x_1, C + \epsilon),$$

First Order Ordinary Differential Equations 61

and therefore on equating terms of order ϵ we have

$$\eta = \xi \left(\frac{\partial S}{\partial x}\right) + \left(\frac{\partial S}{\partial C}\right),$$

where the partial derivatives in brackets refer to y as a function of the two arguments x and C. From this equation and (4.2) we deduce

$$\left(\frac{\partial y}{\partial C}\right) = \eta(x,y) - F(x,y)\xi(x,y). \tag{4.27}$$

Now $\lambda = \eta - F\xi$ satisfies (4.7) and using (4.27) we see that (4.7) could be deduced alternatively in the following two ways.

Firstly, (4.7) follows from differentiating (4.2) partially with respect to C. In the bracket notation for the partial derivatives (4.2) becomes

$$\left(\frac{\partial y}{\partial x}\right) = F(x,y), \tag{4.28}$$

and on partially differentiating with respect to C we obtain,

$$\left(\frac{\partial}{\partial x}\left(\frac{\partial y}{\partial C}\right)\right) = \frac{\partial F}{\partial y}\left(\frac{\partial y}{\partial C}\right).$$

But we have,

$$\left(\frac{\partial}{\partial x}\left(\frac{\partial y}{\partial C}\right)\right) = \frac{\partial}{\partial x}\left(\frac{\partial y}{\partial C}\right) + \frac{\partial}{\partial y}\left(\frac{\partial y}{\partial C}\right)\left(\frac{\partial y}{\partial x}\right),$$

from which (4.7) can be deduced. Secondly, (4.7) follows from the compatibility of the two equations (4.27) and (4.28) which the reader can readily verify.

We see from (4.27) that if $y = y_0(x)$ is a singular solution of (4.2) then $\xi(x,y)$ and $\eta(x,y)$ must be such that

$$\eta(x,y_0) = F(x,y_0)\xi(x,y_0). \tag{4.29}$$

Hence, if as is often the case a singular solution of (4.2) is known, then (4.29) might well suggest the general nature of $\xi(x,y)$ and $\eta(x,y)$. These considerations indicate that singular solutions of first order differential equations perhaps play a more vital role than has been previously considered.

4.6 Invariance of the associated first order partial differential equation

In this section we consider the invariance of the associated first order partial differential equation (4.3). We use the group approach for partial differential equations which is described in detail in subsequent chapters. We look for a one-parameter group of transformations in three variables (x, y, z) which leaves (4.3) invariant. We use the convention that subscripts denote partial differentiation with x, y and z as three independent variables.

Suppose that the one-parameter group

$$x_1 = x + \epsilon \xi(x, y, z) + \mathbf{O}(\epsilon^2),$$
$$y_1 = y + \epsilon \eta(x, y, z) + \mathbf{O}(\epsilon^2), \qquad (4.30)$$
$$z_1 = z + \epsilon \zeta(x, y, z) + \mathbf{O}(\epsilon^2),$$

leaves (4.3) unaltered. We calculate $\frac{\partial z_1}{\partial x_1}$ and $\frac{\partial z_1}{\partial y_1}$ as follows,

$$\frac{\partial z_1}{\partial x_1} = \frac{\partial z_1}{\partial x}\frac{\partial x}{\partial x_1} + \frac{\partial z_1}{\partial y}\frac{\partial y}{\partial x_1}$$
$$= \left\{ \frac{\partial z}{\partial x} + \epsilon \left(\zeta_x + \zeta_z \frac{\partial z}{\partial x} \right) \right\} \left\{ 1 - \epsilon \left(\xi_x + \xi_z \frac{\partial z}{\partial x} \right) \right\}$$
$$+ \left\{ \frac{\partial z}{\partial y} + \epsilon \left(\zeta_y + \zeta_z \frac{\partial z}{\partial y} \right) \right\} \left\{ -\epsilon \left(\eta_x + \eta_z \frac{\partial z}{\partial x} \right) \right\} + \mathbf{O}(\epsilon^2),$$

and therefore

$$\frac{\partial z_1}{\partial x_1} = \frac{\partial z}{\partial x} + \epsilon \left\{ \zeta_x + (\zeta_z - \xi_x)\frac{\partial z}{\partial x} - \eta_x \frac{\partial z}{\partial y} - \xi_z \left(\frac{\partial z}{\partial x}\right)^2 - \eta_z \frac{\partial z}{\partial x}\frac{\partial z}{\partial y} \right\} + \mathbf{O}(\epsilon^2). \qquad (4.31)$$

Similarly,

$$\frac{\partial z_1}{\partial y_1} = \frac{\partial z_1}{\partial x}\frac{\partial x}{\partial y_1} + \frac{\partial z_1}{\partial y}\frac{\partial y}{\partial y_1}$$
$$= \left\{ \frac{\partial z}{\partial x} + \epsilon \left(\zeta_x + \zeta_z \frac{\partial z}{\partial x} \right) \right\} \left\{ -\epsilon \left(\xi_y + \xi_z \frac{\partial z}{\partial y} \right) \right\}$$
$$+ \left\{ \frac{\partial z}{\partial y} + \epsilon \left(\zeta_y + \zeta_z \frac{\partial z}{\partial y} \right) \right\} \left\{ 1 - \epsilon \left(\eta_y + \eta_z \frac{\partial z}{\partial y} \right) \right\} + \mathbf{O}(\epsilon^2),$$

and hence

$$\frac{\partial z_1}{\partial y_1} = \frac{\partial z}{\partial y} + \epsilon \left\{ \zeta_y + (\zeta_z - \eta_y)\frac{\partial z}{\partial y} - \xi_y \frac{\partial z}{\partial x} - \eta_z \left(\frac{\partial z}{\partial y}\right)^2 - \xi_z \frac{\partial z}{\partial x}\frac{\partial z}{\partial y} \right\} + \mathbf{O}(\epsilon^2). \qquad (4.32)$$

First Order Ordinary Differential Equations

If $z = \phi(x, y)$ is a solution of (4.3) then by invariance we have $z_1 = \phi(x_1, y_1)$ and therefore $z = \phi(x, y)$ also satisfies

$$\xi(x, y, z)\frac{\partial z}{\partial x} + \eta(x, y, z)\frac{\partial z}{\partial y} = \zeta(x, y, z). \tag{4.33}$$

Now from,

$$\frac{\partial z_1}{\partial x_1} + F(x_1, y_1)\frac{\partial z_1}{\partial y_1} = 0,$$

and (4.30), (4.31) and (4.32) we can deduce

$$\zeta_x + F\zeta_y = \theta[(\eta - F\xi)_x + F(\eta - F\xi)_y - (\eta - F\xi)F_y], \tag{4.34}$$

where $\theta = \frac{\partial z}{\partial y}$ and we have used $\frac{\partial z}{\partial x} = -F\theta$. From (4.3) and (4.33) we have

$$\zeta = (\eta - \xi F)\theta,$$

and therefore (4.34) gives

$$\frac{\partial \theta}{\partial x} + \frac{\partial}{\partial y}(F\theta) = 0. \tag{4.35}$$

Hence θ, that is $\zeta/(\eta - \xi F)$ is an integrating factor for (4.10). Clearly if z satisfies an equation of the type (4.33) as well as (4.3) then (4.35) can be deduced immediately since from (4.3) and (4.33) we have

$$\frac{\partial z}{\partial x} = \frac{-F\zeta}{(\eta - F\xi)}, \quad \frac{\partial z}{\partial y} = \frac{\zeta}{(\eta - F\xi)},$$

and (4.35) follows from the compatibility of these equations.

In the above we have used the so-called *non-classical* approach for partial differential equations described in a subsequent chapter. We have shown that the first order condition for invariance of (4.3) under (4.30) is equivalent to the existence of an integrating factor for (4.10). Moreover this condition conveys no more information than the condition for the compatibility of (4.3) and (4.33). If we apply the *classical* approach for partial differential equations then on equating coefficients of θ^0 and θ to zero in (4.34) we deduce that $\zeta = \Phi(z)$ and that $\lambda = \eta - F\xi$ satisfies (4.7), with partial derivatives as given in (4.7). Thus the classical approach gives rise to the well known result that if μ is an integrating factor then so also is $\Phi(z)\mu$ where z is the integral of the differential equation.

4.7 Lie's problem and area preserving groups

In this section for a given differential equation (4.2) we attempt to solve (4.6) assuming that the one-parameter group (4.1) is area preserving. That is, we assume there exists a sufficiently continuous and differentiable function $G(x,y)$ such that

$$\xi(x,y) = \frac{\partial G}{\partial y}, \quad \eta(x,y) = -\frac{\partial G}{\partial x}. \tag{4.36}$$

From (4.6) and (4.36) we obtain the second order partial differential equation for $G(x,y)$

$$\frac{\partial^2 G}{\partial x^2} + 2F\frac{\partial^2 G}{\partial x \partial y} + F^2\frac{\partial^2 G}{\partial y^2} = \frac{\partial(G,F)}{\partial(x,y)}, \tag{4.37}$$

which we require to solve for a prescribed function $F(x,y)$. In principle we can solve this equation by introducing two functions $A(G,F)$ and $B(G,F)$ such that

$$\frac{\partial G}{\partial x} = A(G,F), \quad \frac{\partial G}{\partial y} = B(G,F). \tag{4.38}$$

The compatibility condition for $G(x,y)$ together with (4.37) yields two equations for the determination of the first order partial derivatives of $F(x,y)$ and the compatibility condition for this function gives the final equation for $A(G,F)$ and $B(G,F)$. Although the equation obtained is no more tractable than (4.37) the analysis does merit some simplifying features which would seem worthwhile reporting. The following analysis should be contrasted with other possible restrictions concerning the nature of the one-parameter group. For example if $\xi(x,y)$ and $\eta(x,y)$ are assumed to be given as the gradient of some function then this assumption appears to compound the subsequent analysis rather than simplify it. The simplifying features associated with (4.36) may not be due to the fact that the group happens to be area preserving but rather to the fact that (4.36) is embodied in the general expressions for $\xi(x,y)$ and $\eta(x,y)$ (see (2.13)). More precisely $G(x,y)$ is an invariant of the group and (4.36) results from (2.13) in the case when the Jacobian in (2.13) is a function of u only.

In order to solve (4.37) by means of (4.38) we need to assume that the Jacobian

$$J = \frac{\partial(G,F)}{\partial(x,y)}, \tag{4.39}$$

is non-zero and finite. We also need the following elementary relations

$$\begin{aligned}
\frac{\partial x}{\partial G} &= \frac{1}{J}\frac{\partial F}{\partial y}, & \frac{\partial x}{\partial F} &= -\frac{B}{J}, \\
\frac{\partial y}{\partial G} &= -\frac{1}{J}\frac{\partial F}{\partial x}, & \frac{\partial y}{\partial F} &= \frac{A}{J}.
\end{aligned} \tag{4.40}$$

First Order Ordinary Differential Equations 65

Writing the compatibility equation for $G(x,y)$ in the form

$$\frac{\partial(A,x)}{\partial(G,F)} + \frac{\partial(B,y)}{\partial(G,F)} = 0,$$

we obtain

$$\frac{\partial B}{\partial F}\frac{\partial F}{\partial x} - \frac{\partial A}{\partial F}\frac{\partial F}{\partial y} = B\frac{\partial A}{\partial G} - A\frac{\partial B}{\partial G}. \tag{4.41}$$

With $C = A + FB$ we see from (4.7) that (4.37) can be written as

$$\frac{\partial(C,y)}{\partial(G,F)} - F^2\frac{\partial(C/F,x)}{\partial(G,F)} = 0,$$

which on simplification yields,

$$\frac{\partial C}{\partial F}\frac{\partial F}{\partial x} + \left(F\frac{\partial C}{\partial F} - C\right)\frac{\partial F}{\partial y} = -C\frac{\partial C}{\partial G}. \tag{4.42}$$

We note that it is in the derivation of (4.42) that the assumption (4.36) appears to significantly simplify the analysis.

On solving (4.41) and (4.42) for $\partial F/\partial x$ and $\partial F/\partial y$ we obtain

$$\begin{aligned}\frac{\partial F}{\partial x} &= -F\frac{\partial C}{\partial G} - \left(F\frac{\partial C}{\partial F} - C\right)H, \\ \frac{\partial F}{\partial y} &= \frac{\partial C}{\partial G} + \frac{\partial C}{\partial F}H,\end{aligned} \tag{4.43}$$

where $H(G,F)$ is given by

$$H(G,F) = \left\{\frac{\partial C}{\partial G}\frac{\partial C}{\partial F} - C\frac{\partial B}{\partial G}\right\}\left\{\frac{\partial(BC)}{\partial F} - \left(\frac{\partial C}{\partial F}\right)^2\right\}^{-1}. \tag{4.44}$$

We note from (4.38) and (4.43) that the given differential equation (4.2) becomes

$$\frac{dF}{dG} = \left\{\frac{\partial F}{\partial x} + F\frac{\partial F}{\partial y}\right\}\left\{\frac{\partial G}{\partial x} + F\frac{\partial G}{\partial y}\right\}^{-1} = H(G,F). \tag{4.45}$$

From the above equations we find after a long calculation that the compatibility condition for $F(x,y)$ becomes

$$C\frac{\partial^2 C}{\partial G^2} + 2HC\frac{\partial^2 C}{\partial G\partial F} + H^2 C\frac{\partial^2 C}{\partial F^2} + \left(\frac{\partial C}{\partial G}\right)^2 + \left(\frac{\partial C}{\partial F} - B\right)\frac{\partial(HC)}{\partial G} - C\frac{\partial C}{\partial G}\frac{\partial H}{\partial F} = 0, \tag{4.46}$$

which can be written as

$$\frac{\partial}{\partial G}\left\{C\left(\frac{\partial C}{\partial F}-B\right)H+C\frac{\partial C}{\partial G}\right\}+H^2\frac{\partial}{\partial F}\left\{C\left(\frac{\partial C}{\partial F}-B\right)+\frac{C}{H}\frac{\partial C}{\partial G}\right\}=0. \qquad (4.47)$$

On comparing this equation with (4.7) we see that (4.47) is the statement that (4.45) remains invariant under the one-parameter group with infinitesimals $\xi^*(G,F)$ and $\eta^*(G,F)$ given by

$$\xi^*(G,F)=C\left(\frac{\partial C}{\partial F}-B\right), \quad \eta^*(G,F)=-C\frac{\partial C}{\partial G}. \qquad (4.48)$$

Thus an integrating factor for

$$dF-H(G,F)dG=0, \qquad (4.49)$$

is therefore $(H\xi^*-\eta^*)^{-1}$. Now we can verify that

$$H\xi^*-\eta^*=C^2\frac{\partial(A,B)}{\partial(G,F)}\left\{\frac{\partial(BC)}{\partial F}-\left(\frac{\partial C}{\partial F}\right)^2\right\}^{-1}, \qquad (4.50)$$

so that the compatibility condition for $F(x,y)$ reduces to the statement that the differential form

$$\frac{\{C\frac{\partial B}{\partial G}-\frac{\partial C}{\partial G}\frac{\partial C}{\partial F}\}dG+\{\frac{\partial(BC)}{\partial F}-(\frac{\partial C}{\partial F})^2\}dF}{C^2\frac{\partial(A,B)}{\partial(G,F)}}=0, \qquad (4.51)$$

is an exact differential. Problems 13 and 14 illustrate the above analysis with two simple solutions of (4.47). For specific examples we need an expression for the Jacobian J defined by (4.39). From (4.38), (4.39), (4.43) and (4.50) we find that

$$J=C^2\frac{\partial(A,B)}{\partial(G,F)}\left\{\frac{\partial(BC)}{\partial F}-\left(\frac{\partial C}{\partial F}\right)^2\right\}^{-1}. \qquad (4.52)$$

Using $C=A+FB$ we can simplify (4.51) to give

$$\frac{\frac{\partial C}{\partial F}dA+\left(F\frac{\partial C}{\partial F}-C\right)dB}{C^2\frac{\partial(A,B)}{\partial(G,F)}}=0.$$

Thus with $\phi=C^{-1}$ the condition (4.47) is equivalent to the statement that

$$\frac{\partial(\phi,G)}{\partial(A,B)}dA+\frac{\partial(F\phi,G)}{\partial(A,B)}dB=0, \qquad (4.53)$$

First Order Ordinary Differential Equations

is an exact differential. That is the compatibility condition for $F(x,y)$ becomes

$$\frac{\partial\left(\frac{\partial\phi}{\partial B}, G\right)}{\partial(A,B)} + \frac{\partial\left(\phi, \frac{\partial G}{\partial B}\right)}{\partial(A,B)} = \frac{\partial\left(\frac{\partial\psi}{\partial A}, G\right)}{\partial(A,B)} + \frac{\partial\left(\psi, \frac{\partial G}{\partial A}\right)}{\partial(A,B)}, \qquad (4.54)$$

where the functions ϕ and ψ are defined by

$$\phi = \frac{1}{(A+FB)}, \quad \psi = \frac{F}{(A+FB)}. \qquad (4.55)$$

A particular method of solution of (4.54) is outlined in Problems 15, 16, 17 and 18.

It is worthwhile noting that the differential forms (4.51) and (4.53) are consistent with that obtained from the requirement that $(A+FB)^{-1}$ must be an integrating factor for (4.9), *provided* we make use of the expressions (4.43) for $\partial F/\partial x$ and $\partial F/\partial y$. Since from (4.40) and using $C = A+FB$ we have

$$\frac{dy - F\,dx}{(A+FB)} = \frac{C\,dF - \left(\frac{\partial F}{\partial x} + F\frac{\partial F}{\partial y}\right)dG}{JC},$$

and (4.43), (4.44) and (4.52) yields precisely (4.51).

The analysis of the final three sections of this chapter indicate that while it is possible to provide alternative points of view on Lie's problem, it is difficult to make real progress for the general first order differential equation. Lie's problem is a fundamental unsolved problem of mathematics, the solution of which, will no doubt involve an entirely new perspective. Because $\mu = (\eta - F\xi)^{-1}$ is an integrating factor for (4.10), it means that for a given $\mu(x,y)$ there are infinitely many suitable $\xi(x,y)$ and $\eta(x,y)$ and therefore first order differential equations are invariant under infinitely many groups. This is in contrast to higher order differential equations which are invariant under at most a finite number of groups, which is the subject of the next chapter.

PROBLEMS

1. If the differential equation

$$M(x,y)dx + N(x,y)dy = 0,$$

   is invariant under (4.1) show that the infinitesimal condition is equivalent to the existence of an integrating factor $\mu(x,y)$ where

$$\mu(x,y) = \frac{1}{(\xi M + \eta N)}.$$

2. If $\mu(x,y)$ is an integrating factor for both of the differential equations,

$$M(x,y)dx + N(x,y)dy = 0 \text{ and } N(x,y)dx - M(x,y)dy = 0,$$

   show that $\Theta = \tan^{-1}(M/N)$ satisfies

$$\nabla^2 \Theta = \frac{\partial^2 \Theta}{\partial x^2} + \frac{\partial^2 \Theta}{\partial y^2} = 0.$$

3. For the quasi-linear first order partial differential equation

$$a(x,y,z)\frac{\partial z}{\partial x} + b(x,y,z)\frac{\partial z}{\partial y} = c(x,y,z), \qquad (*)$$

   show that the general solution is given by

$$\rho = \Phi(\sigma),$$

   where Φ is an arbitrary function and $\rho(x,y,z)$ and $\sigma(x,y,z)$ are any two independent integrals of the system of differential equations

$$\frac{dx}{d\tau} = a(x,y,z), \quad \frac{dy}{d\tau} = b(x,y,z), \quad \frac{dz}{d\tau} = c(x,y,z).$$

   [Hint, from $\rho = $ constant and $\sigma = $ constant we have

$$\frac{d\rho}{d\tau} = a\rho_x + b\rho_y + c\rho_z = 0,$$

$$\frac{d\sigma}{d\tau} = a\sigma_x + b\sigma_y + c\sigma_z = 0,$$

First Order Ordinary Differential Equations

where subscripts denote partial differentiation with x, y and z as three independent variables. These two equations together with (*) constitute three homogeneous equations for a, b and c. For non-trivial solutions the determinant vanishes and this condition can be shown to become

$$\frac{\partial(\rho, \sigma)}{\partial(x, y)} = 0,$$

from which the required condition follows. In the Jacobian partial derivatives are with x and y as the independent variables, that is

$$\frac{\partial \rho}{\partial x} = \rho_x + \rho_z \frac{\partial z}{\partial x}, \quad \frac{\partial \sigma}{\partial y} = \sigma_y + \sigma_z \frac{\partial z}{\partial y}, \quad \text{etc.}]$$

4. If $y_0(x)$ is a known solution of the Riccati equation

$$\frac{dy}{dx} + p(x)y = q(x) + r(x)y^2,$$

show that the substitution $y = y_0 + w^{-1}$ gives rise to the linear equation

$$\frac{dw}{dx} + [2r(x)y_0(x) - p(x)]w = -r(x).$$

5. Show that the most general first order differential equation which admits the group

$$\xi(x, y) = \xi(x), \quad \eta(x, y) = \eta(x)y + \zeta(x),$$

is

$$\frac{dy}{dx} - \frac{\eta(x)}{\xi(x)}y - \frac{\zeta(x)}{\xi(x)} = \frac{1}{\xi(x)s(x)} \Phi\left(s(x)y - \int_{x_0}^{x} \frac{\zeta(t)}{\xi(t)} s(t) dt\right),$$

where $s(x)$ is given by

$$s(x) = e^{-\int_{x_0}^{x} \frac{\eta(t)}{\xi(t)} dt},$$

and Φ is an arbitrary function of the argument indicated.

6. Using (4.15)$_2$ and (4.18), make the substitution $w = \xi W$ and deduce the equation,

$$\left(\frac{\eta}{\xi}\right) \frac{dW}{dy} + \frac{\partial}{\partial y}\left(\frac{\eta}{\xi}\right) W = \frac{1}{\xi^2} \frac{\partial \xi}{\partial y}.$$

7. Show that the most general first order differential equation invariant under the group

$$\xi(x,y) = \xi(x)y^{n-1}, \quad \eta(x,y) = \eta(x)y^n,$$

is

$$y\left(\frac{dy}{dx} - \frac{\eta(x)}{\xi(x)}y\right)^{-1} = \frac{\xi(x)}{s(x)^{n-1}}\left\{(n-1)\int_{x_0}^{x}\frac{s(t)^{n-1}}{\xi(t)}dt + \Phi[s(x)y]\right\},$$

where Φ is an arbitrary function and $s(x)$ is as defined in Problem 5. Using canonical coordinates

$$u(x,y) = s(x)y, \quad v(x,y) = \frac{1}{y^{n-1}s(x)^{n-1}}\int_{x_0}^{x}\frac{s(t)^{n-1}}{\xi(t)}dt,$$

show that the differential equation becomes

$$\frac{du}{dv} = \frac{u^n}{\Phi(u)}.$$

8. For the Riccati equation given in Problem 4, show that the substitution

$$y(x) = -\frac{1}{r(x)z(x)}\frac{dz}{dx},$$

gives rise to the linear equation,

$$\frac{d^2z}{dx^2} + \left(p(x) - \frac{1}{r(x)}\frac{dr}{dx}\right)\frac{dz}{dx} + q(x)r(x)z = 0.$$

Deduce the normal form of this differential equation.

9. **Continuation.** Show that Riccati equation of Problem 4, admits the group

$$\xi(x,y) = \xi(x), \quad \eta(x,y) = \eta(x)y + \zeta(x),$$

provided,

$$\frac{d}{dx}(r\xi) = -r\eta, \quad \frac{d}{dx}(p\xi + \eta) = 2r\zeta,$$

and

$$\frac{d}{dx}(q\xi - \zeta) = p\zeta + q\eta.$$

First Order Ordinary Differential Equations

From these equations deduce that $\xi(x)$ satisfies the equation,

$$(2\xi\xi'' - \xi'^2) + 4\left\{qr - \frac{1}{2}\frac{d}{dx}\left(p - \frac{1}{2}\frac{dr}{dx}\right) - \frac{1}{4}\left(p - \frac{1}{r}\frac{dr}{dx}\right)^2\right\}\xi^2 = C,$$

where C is a constant. Can you reconcile this result with that of Section 3.3.

[The following three problems summarize the three criteria given by Dickson (1924) (page 313) for the invariance of a differential equation under a one-parameter group.]

10. Show that a first order ordinary differential equation is invariant under the group

$$x_1 = x + \epsilon\xi(x,y) + \mathbf{O}(\epsilon^2), \quad y_1 = y + \epsilon\eta(x,y) + \mathbf{O}(\epsilon^2), \qquad (**)$$

if and only if

$$Lz = \Phi(z),$$

where $z(x,y)$ is the integral of the equation, L is the operator

$$L = \xi\frac{\partial}{\partial x} + \eta\frac{\partial}{\partial y},$$

and Φ denotes an arbitrary function.

11. **Continuation.** The differential operator associated with

$$M(x,y)dx + N(x,y)dy = 0, \qquad (+)$$

is given by

$$P = N\frac{\partial}{\partial x} - M\frac{\partial}{\partial y}.$$

The commutator (LP) is defined by

$$(LP) = LP - PL.$$

Show that,

(i) $$(LP) = (LN - P\xi)\frac{\partial}{\partial x} - (LM + P\eta)\frac{\partial}{\partial y}.$$

(ii) the differential equation $(+)$ is invariant under $(**)$ if and only if the commutator (LP) is a constant multiple of the operator P.

12. **Continuation.** The first extension of the operator L is L' where

$$L' = \xi \frac{\partial}{\partial x} + \eta \frac{\partial}{\partial y} + \pi \frac{\partial}{\partial y'},$$

and where π is given by

$$\pi = \frac{\partial \eta}{\partial x} + \left(\frac{\partial \eta}{\partial y} - \frac{\partial \xi}{\partial x}\right) y' - \frac{\partial \xi}{\partial y} y'^2.$$

Show that the first order differential equation

$$F(x, y, y') = 0,$$

remains invariant under (**) if and only if

$$L'F = 0.$$

[The following six problems relate to Section 4.7.]

13. Assuming that

$$A = f(G), \quad B = g(G)F^{-1},$$

where f and g are functions of G only, show that equation (4.47) simplifies to yield

$$f''(1 + f/g) + f'(f/g)' = 0,$$

where primes denote differentiation with respect to G. Integrate this equation and show that $J = \alpha g$ where α is the integration constant. Hence from the relations (4.40) deduce that

$$x = -\frac{1}{\alpha} \log\left(\frac{fF}{g}\right) + \int_{G_0}^{G} \frac{dt}{f(t)}, \quad y = \frac{fF}{\alpha g} - \frac{\beta}{\alpha},$$

where β and G_0 are further integration constants. From these results show that the original differential equation (4.2) in this case has the form

$$\frac{dy}{dx} = (\alpha y + \beta) h[\alpha x + \log(\alpha y + \beta)],$$

for some arbitrary function h of the argument indicated. Observe that this equation can be solved by the substitution

$$\rho = \alpha x + \log(\alpha y + \beta).$$

14. Assuming that $H(G, F)$ is identically zero show from (4.44) and (4.47) that

$$B = \frac{\{\ell'(F)[\ell(F)G + m(F)] - [\ell(F)m'(F) - m(F)\ell'(F)]\}}{2\ell(F)[\ell(F)G + m(F)]^{1/2}} + n(F),$$

$$C = [\ell(F)G + m(F)]^{1/2},$$

where ℓ, m and n denote arbitrary functions of F and here primes denote differentiation with respect to F. Show that $J = \ell(F)/2$ and from the relations (4.40) deduce

$$x = \frac{2}{\ell(F)}[\ell(F)G + m(F)]^{1/2} + p(F), \quad y = \frac{2F}{\ell(F)}[\ell(F)G + m(F)]^{1/2} + q(F),$$

for some functions $p(F)$ and $q(F)$. With $s(F) = q(F) - Fp(F)$ show that in this case the original differential equation (4.2) is the well known Clairaut's equation (Murphy (1960), page 65)

$$y = x\frac{dy}{dx} + s\left(\frac{dy}{dx}\right),$$

which has general solution $y = \gamma x + s(\gamma)$ for some constant γ.

15. Assuming there exists some function $w(A, B)$ such that ϕ and ψ as defined by (4.55) are given by

$$\phi = \frac{\partial w}{\partial B}, \quad \psi = -\frac{\partial w}{\partial A}, \tag{$*$}$$

show that equation (4.54) becomes

$$\frac{\partial(\nabla^2 w, G)}{\partial(A, B)} + \frac{\partial\left(\frac{\partial w}{\partial A}, \frac{\partial G}{\partial A}\right)}{\partial(A, B)} + \frac{\partial\left(\frac{\partial w}{\partial B}, \frac{\partial G}{\partial B}\right)}{\partial(A, B)} = 0, \tag{$**$}$$

where the Laplacian ∇^2 is given by

$$\nabla^2 = \frac{\partial^2}{\partial A^2} + \frac{\partial^2}{\partial B^2}.$$

Show that $(**)$ can be written alternatively as

$$\nabla^2 \frac{\partial(w, G)}{\partial(A, B)} = \frac{\partial(w, \nabla^2 G)}{\partial(A, B)} - \frac{\partial(\nabla^2 w, G)}{\partial(A, B)}. \tag{$***$}$$

16. **Continuation.** From (4.55) and $(*)$ conclude that $w(A, B)$ satisfies the first order partial differential equation

$$A\frac{\partial w}{\partial B} - B\frac{\partial w}{\partial A} = 1,$$

and hence

$$w = \tan^{-1}(B/A) + f[(A^2 + B^2)^{1/2}],$$

where f denotes an arbitrary function of the argument indicated. Introducing polar coordinates

$$R = (A^2 + B^2)^{1/2}, \quad \Theta = \tan^{-1}(B/A),$$

show that

$$F = [B - Ag(R)]/[A + Bg(R)],$$

where $g(R) = Rf'(R)$ and the prime denotes differentiation with respect to R.

17. **Continuation.** In (R, Θ) coordinates observe that $(***)$ of Problem 15 becomes

$$\nabla^2 \frac{1}{R} \frac{\partial(w, G)}{\partial(R, \Theta)} = \frac{1}{R} \left\{ \frac{\partial(w, \nabla^2 G)}{\partial(R, \Theta)} - \frac{\partial(\nabla^2 w, G)}{\partial(R, \Theta)} \right\}, \qquad (****)$$

where ∇^2 is given by

$$\nabla^2 = \frac{\partial^2}{\partial R^2} + \frac{1}{R} \frac{\partial}{\partial R} + \frac{1}{R^2} \frac{\partial^2}{\partial \Theta^2}.$$

On using $w = \Theta + f(R)$ show that $(****)$ simplifies to give

$$\frac{\partial^2 G}{\partial R^2} - \frac{1}{R} \frac{\partial G}{\partial R} - \frac{1}{R^2} \frac{\partial^2 G}{\partial \Theta^2} + \frac{\partial}{\partial R} \left\{ \left(g' - \frac{2g}{R} \right) \frac{\partial G}{\partial \Theta} \right\} = 0,$$

where $g(R)$ is as defined in Problem 16.

18. **Continuation.** With $G = R^2$ and $h(G) \equiv g(G^{1/2})$ show that

$$A = \frac{G^{1/2}[1 - Fh(G)]}{\{[1 + F^2][1 + h(G)^2]\}^{1/2}}, \quad B = \frac{G^{1/2}[F + h(G)]}{\{[1 + F^2][1 + h(G)^2]\}^{1/2}}.$$

Hence show that $J = (1 + F^2)/2$ and that the relations (4.40) yield, apart from arbitrary additive constants

$$x = \frac{2G^{1/2}[1 - Fh(G)]}{\{[1 + F^2][1 + h(G)^2]\}^{1/2}}, \quad y = \frac{2G^{1/2}[F + h(G)]}{\{[1 + F^2][1 + h(G)^2]\}^{1/2}}.$$

First Order Ordinary Differential Equations

Hence conclude that the original differential equation (4.2) in this case is

$$\frac{dy}{dx} = \left\{\frac{y - xg[(x^2+y^2)^{1/2}/2]}{x + yg[(x^2+y^2)^{1/2}/2]}\right\},$$

which is solved using polar coordinates (see Murphy (1960), page 67).

19. Given the one-parameter group

$$x_1 = x + \epsilon\xi(x,y) + \mathbf{O}(\epsilon^2), \quad y_1 = y + \epsilon\eta(x,y) + \mathbf{O}(\epsilon^2),$$

show that

$$y_1(x_1 - x_0) = y(x - x_0)$$
$$+ \epsilon\left\{[\xi(x,y) - \xi(x - x_0, y(x - x_0))]\frac{dy}{dx}(x - x_0) + \eta(x - x_0, y(x - x_0))\right\} + \mathbf{O}(\epsilon^2).$$

[This result can be verified by two distinct methods.

(i) Suppose that $y = S(x,C)$ and $y_1 = S(x_1, C + \epsilon)$ then $y(x - x_0) = S(x - x_0, C)$ and

$$y_1(x_1 - x_0) = S(x_1 - x_0, C + \epsilon),$$
$$= S(x - x_0 + \epsilon\xi(x,y), C + \epsilon) + \mathbf{O}(\epsilon^2),$$
$$= S(x - x_0, C) + \epsilon\left\{\xi(x,y)\left(\frac{\partial S}{\partial x}\right) + \left(\frac{\partial S}{\partial C}\right)\right\} + \mathbf{O}(\epsilon^2),$$
$$= y(x - x_0) + \epsilon\left\{\xi(x,y)\frac{dy}{dx}(x - x_0) + \left(\frac{\partial S}{\partial C}\right)\right\} + \mathbf{O}(\epsilon^2),$$

where the partial derivative of S with respect to C has arguments $x - x_0$ and C and is found from (4.27) to be given by

$$\left(\frac{\partial S}{\partial C}\right) = \eta(x - x_0, y(x - x_0)) - \xi(x - x_0, y(x - x_0))\frac{dy}{dx}(x - x_0),$$

from which the required result follows.

(ii) Alternatively we have

$$y(x - x_0) = e^{-x_0 \, d/dx}y(x),$$

and we require to find

$$y_1(x_1 - x_0) = e^{-x_0 \, d/dx_1}y_1(x_1).$$

From

$$x = x_1 - \epsilon\xi(x_1, y_1) + \mathbf{O}(\epsilon^2),$$

we have

$$dx = dx_1\left\{1 - \epsilon\left(\frac{\partial\xi}{\partial x} + \frac{\partial\xi}{\partial y}\frac{dy}{dx}\right)\right\} + \mathbf{O}(\epsilon^2),$$

and therefore

$$\frac{d}{dx_1} = \left\{1 - \epsilon\left(\frac{\partial\xi}{\partial x} + \frac{\partial\xi}{\partial y}\frac{dy}{dx}\right)\right\}\frac{d}{dx} + \mathbf{O}(\epsilon^2).$$

Hence if we define the differential operators D_1 and D_2 by

$$D_1 = \frac{d}{dx}, \quad D_2 = -\left(\frac{\partial\xi}{\partial x} + \frac{\partial\xi}{\partial y}\frac{dy}{dx}\right)\frac{d}{dx},$$

then we require to evaluate

$$y_1(x_1 - x_0) = e^{-x_0(D_1 + \epsilon D_2)}[y + \epsilon\eta(x, y)] + \mathbf{O}(\epsilon^2).$$

Gröbner and Knapp (1967)(page 40) give formulae for operators of this type. Observe that,

$$e^{-x_0 D_1}x = x - x_0, \quad e^{-x_0 D_1}y = y(x - x_0),$$

and that

$$D_2 y(x + \tau) = -\frac{d\xi}{dx}(x, y)\frac{dy}{dx}(x + \tau).$$

In order to calculate the order of ϵ term arising from

$$e^{-x_0(D_1 + \epsilon D_2)}y,$$

we use the integral given in Gröbner and Knapp (1967)(page 40). We have

$$e^{-x_0(D_1 + \epsilon D_2)}y = y(x - x_0) + \epsilon\int_0^{-x_0}[D_2 y(x + \tau)]^* \, d\tau + \mathbf{O}(\epsilon^2),$$

where the 'star' in the integrand denotes that (x, y) in the square bracket becomes $(x - x_0 - \tau, y(x - x_0 - \tau))$. If in the integral we make the substitution

$$\rho = x - x_0 - \tau,$$

First Order Ordinary Differential Equations

then we have

$$e^{-x_0(D_1+\epsilon D_2)}y = y(x-x_0) + \epsilon \int_{x-x_0}^{x} \frac{d\xi}{d\rho}(\rho, y(\rho))\frac{dy}{dx}(x-x_0)d\rho + \mathbf{O}(\epsilon^2),$$

and thus

$$e^{-x_0(D_1+\epsilon D_2)}y = y(x-x_0) + \epsilon\frac{dy}{dx}(x-x_0)\left[\xi(x,y) - \xi(x-x_0,\ y(x-x_0))\right] + \mathbf{O}(\epsilon^2).$$

The result now follows since

$$e^{-x_0(D_1+\epsilon D_2)}\epsilon\eta(x,y) = \epsilon\eta(x-x_0,\ y(x-x_0)) + \mathbf{O}(\epsilon^2).$$

Note that we can check the validity of this second method by using the integral given in Gröbner and Knapp (1967) to evaluate

$$x_1 - x_0 = e^{-x_0(D_1+\epsilon D_2)}[x + \epsilon\xi(x,y)] + \mathbf{O}(\epsilon^2).$$

Proceeding as above we obtain

$$x_1 - x_0 = x - x_0 + \epsilon\xi(x,y) + \mathbf{O}(\epsilon^2),$$

which of course is the desired result.]

20. **Continuation.** Show that the one-parameter groups

$$x_1 = x + \epsilon\xi(x) + \mathbf{O}(\epsilon^2), \quad y_1 = y + \epsilon\eta(x)y + \mathbf{O}(\epsilon^2),$$

which leave the following differential-difference equations invariant,

(i) $$\frac{dy}{dx}(x) = -y(x-x_0),$$

(ii) $$\frac{dy}{dx}(x) = y(x)[1 - y(x-x_0)],$$

(iii) $$\frac{dy}{dx}(x) = y(x)[y(x) - y(x-x_0)],$$

are respectively as follows,

(i) $$x_1 = x + \alpha\epsilon, \quad y_1 = e^{\beta\epsilon}y,$$

(ii) $$x_1 = x + \epsilon, \quad y_1 = y,$$

(iii) $$x_1 = e^{-\alpha\epsilon}x + \beta\alpha^{-1}(1 - e^{-\alpha\epsilon}), \quad y_1 = e^{\alpha\epsilon}y,$$

where α and β are arbitrary constants. Can we use these groups to simplify or integrate the equation?

[Notice, that since

$$y(x - x_0) = e^{-x_0\, d/dx}y(x) = \sum_{n=0}^{\infty} \frac{(-x_0)^n}{n!} y^{(n)}(x),$$

differential-difference equations are really 'infinite' order differential equations.]

Chapter Five
Second and higher order ordinary differential equations

5.1 Introduction

First order differential equations can be invariant under an infinite number of one-parameter groups. Second and higher order equations differ in that they are invariant under at most a finite number of groups. Second order equations are invariant under at most eight while for $n > 2$, n^{th} order differential equations are invariant under at most $n+4$ groups (see Dickson (1924), page 353). Higher order equations also differ from first order ones in that if there exists a one-parameter group leaving the equation invariant then this group can be systematically determined. Much of the literature is concerned with obtaining the most general second order differential equation invariant under a given group and again we advise the reader to consult standard tables of such differential equations (see for example Dickson (1924), page 349). In the following section we deduce the condition (5.8) for a second order differential equation to be invariant under a one-parameter group and in the next section we give four examples making use of this condition. In the section thereafter we give examples of the determination of the most general second and higher order differential equations invariant under a given one-parameter group and in the final section we give three applications from the nuclear industry.

5.2 Infinitesimal versions of y'' and $y'' = F(x, y, y')$

We consider the general second order differential equation

$$\frac{d^2y}{dx^2} = F\left(x, y, \frac{dy}{dx}\right), \tag{5.1}$$

and look for a one-parameter group,

$$x_1 = x + \epsilon\xi(x,y) + \mathbf{O}(\epsilon^2), \quad y_1 = y + \epsilon\eta(x,y) + \mathbf{O}(\epsilon^2), \tag{5.2}$$

which leaves (5.1) invariant. Throughout this chapter we shall use the notation

$$z = \frac{dy}{dx}, \tag{5.3}$$

so that from results given in the previous chapter we have

$$z_1 = z + \epsilon\pi(x,y,z) + \mathbf{O}(\epsilon^2), \tag{5.4}$$

where $\pi(x,y,z)$ is given by

$$\pi = \frac{\partial \eta}{\partial x} + \left(\frac{\partial \eta}{\partial y} - \frac{\partial \xi}{\partial x}\right)z - \frac{\partial \xi}{\partial y}z^2. \tag{5.5}$$

In order to calculate the infinitesimal version of y'' we proceed as follows,

$$\frac{d^2 y_1}{dx_1^2} = \frac{d}{dx}\left(\frac{dy_1}{dx_1}\right)\frac{dx}{dx_1}$$

$$= \left\{\frac{d^2 y}{dx^2} + \epsilon\left(\frac{\partial \pi}{\partial x} + \frac{\partial \pi}{\partial y}\frac{dy}{dx} + \frac{\partial \pi}{\partial z}\frac{dz}{dx}\right)\right\}\left\{1 - \epsilon\left(\frac{\partial \xi}{\partial x} + \frac{\partial \xi}{\partial y}\frac{dy}{dx}\right)\right\} + \mathbf{O}(\epsilon^2),$$

and thus we have

$$\frac{d^2 y_1}{dx_1^2} = \frac{d^2 y}{dx^2} + \epsilon\left\{\left(\frac{\partial \pi}{\partial x} + \frac{\partial \pi}{\partial y}z + \frac{\partial \pi}{\partial z}\frac{d^2 y}{dx^2}\right) - \left(\frac{\partial \xi}{\partial x} + \frac{\partial \xi}{\partial y}z\right)\frac{d^2 y}{dx^2}\right\} + \mathbf{O}(\epsilon^2). \tag{5.6}$$

If (5.1) is left invariant by (5.2) then on using

$$F(x_1, y_1, z_1) = F(x, y, z) + \epsilon\left(\xi\frac{\partial F}{\partial x} + \eta\frac{\partial F}{\partial y} + \pi\frac{\partial F}{\partial z}\right) + \mathbf{O}(\epsilon^2),$$

and from (5.1) and (5.6) we deduce the condition that (5.2) leaves (5.1) invariant, namely

$$\left(\frac{\partial \pi}{\partial x} + \frac{\partial \pi}{\partial y}z + \frac{\partial \pi}{\partial z}F\right) - \left(\frac{\partial \xi}{\partial x} + \frac{\partial \xi}{\partial y}z\right)F = \left(\xi\frac{\partial F}{\partial x} + \eta\frac{\partial F}{\partial y} + \pi\frac{\partial F}{\partial z}\right). \tag{5.7}$$

If F involves powers of z then generally we can determine ξ and η from (5.7) by equating coefficients of the powers of z. On using (5.5) we find that (5.7) becomes

$$\left\{\frac{\partial^2 \eta}{\partial x^2} + \left(\frac{\partial \eta}{\partial y} - 2\frac{\partial \xi}{\partial x}\right)F - \xi\frac{\partial F}{\partial x} - \eta\frac{\partial F}{\partial y} - \frac{\partial \eta}{\partial x}\frac{\partial F}{\partial z}\right\}$$

$$+ \left\{2\frac{\partial^2 \eta}{\partial x \partial y} - \frac{\partial^2 \xi}{\partial x^2} - 3F\frac{\partial \xi}{\partial y} - \left(\frac{\partial \eta}{\partial y} - \frac{\partial \xi}{\partial x}\right)\frac{\partial F}{\partial z}\right\}z$$

$$+ \left\{\frac{\partial^2 \eta}{\partial y^2} - 2\frac{\partial^2 \xi}{\partial x \partial y} + \frac{\partial \xi}{\partial y}\frac{\partial F}{\partial z}\right\}z^2 - \frac{\partial^2 \xi}{\partial y^2}z^3$$

$$= 0.$$

$$\tag{5.8}$$

In the following section we illustrate with examples how solutions $\xi(x,y)$ and $\eta(x,y)$ of (5.8) can be deduced.

5.3 Examples of the determination of $\xi(x,y)$ and $\eta(x,y)$

Example 5.1 Show that if $F(x,y,y')$ is independent of y' then $\xi(x,y)$ and $\eta(x,y)$ take the following forms,

$$\xi(x,y) = \rho(x)y + \xi(x), \quad \eta(x,y) = \rho'(x)y^2 + \eta(x)y + \zeta(x). \tag{5.9}$$

If $F(x,y,z)$ is independent of z then (5.8) becomes

$$\begin{aligned}
&\left\{\frac{\partial^2 \eta}{\partial x^2} + \left(\frac{\partial \eta}{\partial y} - 2\frac{\partial \xi}{\partial x}\right)F - \xi\frac{\partial F}{\partial x} - \eta\frac{\partial F}{\partial y}\right\} \\
&+ \left\{2\frac{\partial^2 \eta}{\partial x \partial y} - \frac{\partial^2 \xi}{\partial x^2} - 3F\frac{\partial \xi}{\partial y}\right\}z \\
&+ \left\{\frac{\partial^2 \eta}{\partial y^2} - 2\frac{\partial^2 \xi}{\partial x \partial y}\right\}z^2 - \frac{\partial^2 \xi}{\partial y^2}z^3 \\
&= 0,
\end{aligned} \tag{5.10}$$

and from the coefficients of z^2 and z^3 we have

$$\frac{\partial^2 \eta}{\partial y^2} - 2\frac{\partial^2 \xi}{\partial x \partial y} = 0, \quad \frac{\partial^2 \xi}{\partial y^2} = 0,$$

and (5.9) follows immediately. We notice that from the coefficients of z and z^0 we also have

$$3\rho(x)F = 3\rho''(x)y + [2\eta'(x) - \xi''(x)],$$

$$[\rho(x)y + \xi(x)]\frac{\partial F}{\partial x} + [\rho'(x)y^2 + \eta(x)y + \zeta(x)]\frac{\partial F}{\partial y} \tag{5.11}$$

$$= [\eta(x) - 2\xi'(x)]F + [\rho'''(x)y^2 + \eta''(x)y + \zeta''(x)].$$

Hence either $\rho(x)$ is identically zero and $2\eta'(x) = \xi''(x)$ and $F(x,y)$ is a solution of the partial differential equation

$$\xi(x)\frac{\partial F}{\partial x} + [\eta(x)y + \zeta(x)]\frac{\partial F}{\partial y} = [\eta(x) - 2\xi'(x)]F + [\eta''(x)y + \zeta''(x)],$$

or $\rho(x)$ is non-zero in which case $F(x,y)$ must be a linear function of y (see Problems 1 and 2).

Example 5.2 Show that $y'' = 0$ is invariant under precisely 8 one-parameter groups.

From (5.8) with F identically zero we have

$$\frac{\partial^2 \eta}{\partial x^2} + \left(2\frac{\partial^2 \eta}{\partial x \partial y} - \frac{\partial^2 \xi}{\partial x^2}\right)z + \left(\frac{\partial^2 \eta}{\partial y^2} - 2\frac{\partial^2 \xi}{\partial x \partial y}\right)z^2 - \frac{\partial^2 \xi}{\partial y^2}z^3 = 0, \tag{5.12}$$

and as in the previous example we deduce that $\xi(x,y)$ and $\eta(x,y)$ must be of the form (5.9). From the coefficients of z and z^0 in (5.12) and (5.9) we deduce

$$\xi''(x) = 2\eta'(x), \quad \rho''(x) = \eta''(x) = \zeta''(x) = 0,$$

and hence

$$\rho(x) = C_1 x + C_2, \quad \xi(x) = C_3 x^2 + C_4 x + C_5,$$
$$\eta(x) = C_3 x + C_6, \quad \zeta(x) = C_7 x + C_8,$$

where $C_1, C_2, \ldots, C_8$ denote arbitrary constants. Each of these constants gives rise to a one-parameter group leaving $y'' = 0$ invariant. Consider for example the group generated by C_3, that is take $C_3 = 1$ and assume all the other constants are zero. From

$$\frac{dx_1}{d\epsilon} = x_1^2, \quad \frac{dy_1}{d\epsilon} = x_1 y_1,$$

and $x_1 = x$, $y_1 = y$ when $\epsilon = 0$ we find that the global form of the group is

$$x_1 = \frac{x}{(1 - \epsilon x)}, \quad y_1 = \frac{y}{(1 - \epsilon x)}.$$

We have

$$\frac{dy_1}{dx_1} = (1 - \epsilon x)\frac{dy}{dx} + \epsilon y, \quad \frac{d^2 y_1}{dx_1^2} = (1 - \epsilon x)^3 \frac{d^2 y}{dx^2},$$

so that clearly $y'' = 0$ remains invariant under this group.

Example 5.3 Show that the differential equation

$$\frac{d^2 y}{dx^2} = xy + \exp\left(\frac{dy}{dx}\right),$$

is not invariant under any one-parameter group.

With $F = xy + e^z$ we see that (5.8) becomes

$$\left\{\frac{\partial^2 \eta}{\partial x^2} + xy\left(\frac{\partial \eta}{\partial y} - 2\frac{\partial \xi}{\partial x}\right) - y\xi - x\eta\right\} + \left\{2\frac{\partial^2 \eta}{\partial x \partial y} - \frac{\partial^2 \xi}{\partial x^2} - 3xy\frac{\partial \xi}{\partial y}\right\} z$$
$$+ \left\{\frac{\partial^2 \eta}{\partial y^2} - 2\frac{\partial^2 \xi}{\partial x \partial y}\right\} z^2 - \frac{\partial^2 \xi}{\partial y^2} z^3$$
$$= e^z \left\{\left(2\frac{\partial \xi}{\partial x} - \frac{\partial \eta}{\partial y} + \frac{\partial \eta}{\partial x}\right) + \left(3\frac{\partial \xi}{\partial y} + \frac{\partial \eta}{\partial y} - \frac{\partial \xi}{\partial x}\right) z - \frac{\partial \xi}{\partial y} z^2\right\}.$$

Second and Higher Order Ordinary Differential Equations

From the term involving e^z we deduce

$$\frac{\partial \xi}{\partial y} = 0, \quad \frac{\partial \xi}{\partial x} = \frac{\partial \eta}{\partial y} = -\frac{\partial \eta}{\partial x},$$

so that $\xi(x,y) = C_1 x + C_2$ and $\eta(x,y) = C_1(y-x) + C_3$ where C_1, C_2 and C_3 denote arbitrary constants. From the coefficient of z^0 in the term not involving e^z we have

$$-C_1 xy = y(C_1 x + C_2) + x[C_1(y-x) + C_3],$$

which is clearly only satisfied if $C_1 = C_2 = C_3 = 0$. Hence there is no one-parameter group leaving the given differential equation invariant.

Example 5.4 Obtain the most general invariant one-parameter group for the second order differential equation

$$\frac{d^2 y}{dx^2} + p(x)y = 0. \tag{5.13}$$

As in Example 5.1 $\xi(x,y)$ and $\eta(x,y)$ are given by (5.9) and from (5.11) with $F(x,y) = -p(x)y$ we deduce

$$\rho''(x) + p(x)\rho(x) = 0,$$

$$2\eta'(x) = \xi''(x), \quad \eta''(x) + p'(x)\xi(x) + 2p(x)\xi'(x) = 0, \tag{5.14}$$

$$\zeta''(x) + p(x)\zeta(x) = 0,$$

and hence the given differential equation is invariant under 8 distinct one-parameter groups. We notice that the group arising from $\xi(x)$ and $\eta(x)$ is that considered in Section 3.3. The group arising from $\zeta(x)$ merely reflects the invariance of (5.13) under the addition to y of any solution of (5.13). We consider $\rho(x)$ in more detail. For this group we find that suitable canonical coordinates are

$$u(x,y) = \frac{y}{\rho(x)}, \quad v(x,y) = \frac{\rho(x)}{y}\int_{x_0}^{x} \frac{dt}{\rho(t)^2},$$

and we have

$$\frac{du}{d(uv)} = \rho(x)\frac{dy}{dx} - \rho'(x)y.$$

Hence on using (5.13) and (5.14)$_1$ we obtain

$$\frac{d}{dx}\left(\frac{du}{d(uv)}\right) = \rho(x)\frac{d^2 y}{dx^2} - \rho''(x)y = 0,$$

and therefore we have

$$u = C_1 uv + C_2,$$

where C_1 and C_2 are constants. From this equation we readily deduce

$$y = C_1 \rho(x) \int_{x_0}^{x} \frac{dt}{\rho(t)^2} + C_2 \rho(x),$$

which is a well known result for the general solution of (5.13).

It is also worthwhile noting that embodied in $(5.14)_2$ ($\xi(x) = 0$, $\eta(x) = 1$) is the group

$$x_1 = x, \quad y_1 = e^\epsilon y,$$

which reflects the invariance of all linear homogeneous equations under stretchings of y. In this case suitable canonical coordinates are

$$u(x, y) = x, \quad v(x, y) = \log y,$$

and with $w(u)$ defined by

$$w(u) = \frac{dv}{du},$$

the differential equation (5.13) becomes the first order Riccati equation (Murphy (1960), page 15),

$$\frac{dw}{du} + w^2 + p(u) = 0.$$

5.4 Determination of the most general differential equation invariant under a given group

In the notation of Section 5.2 suppose for given $\xi(x,y)$ and $\eta(x,y)$ we have deduced two independent invariants $A(x,y)$ and $B(x,y,z)$ (say) of the characteristic equations

$$\frac{dx}{d\tau} = \xi(x,y), \quad \frac{dy}{d\tau} = \eta(x,y), \quad \frac{dz}{d\tau} = \pi(x,y,z), \tag{5.15}$$

where $\pi(x,y,z)$ is defined by (5.5). The basic result for the determination of the most general second order equation invariant under this group is that this equation is given by

$$\frac{dB}{dA} = \Phi(A, B), \tag{5.16}$$

where Φ is an arbitrary function of the arguments indicated. Clearly (5.16) is of second order and is invariant under the given group. In order to see that there can be no more general equation than (5.16) we refer the reader to the comment following Problem 5.

We remark also that the most general third order differential equation invariant under the given group is given by

$$\frac{d^2 B}{dA^2} = \Psi\left(A,\ B,\ \frac{dB}{dA}\right), \tag{5.17}$$

where Ψ denotes an arbitrary function.

Example 5.5 Obtain the most general second order differential equation invariant under the group

$$x_1 = f(x), \quad y_1 = g(x)y,$$

and hence deduce the most general linear invariant second order equation.

From Example 4.3 we have already obtained

$$A = s(x)y, \quad B = s(x)[\xi(x)z - \eta(x)y], \tag{5.18}$$

where $s(x)$, $\xi(x)$ and $\eta(x)$ are as previously defined. Now

$$\frac{dB}{dA} = \frac{\left\{\xi(x)\frac{d^2 y}{dx^2} + (\xi'(x) - 2\eta(x))\frac{dy}{dx} + \left(\frac{\eta(x)^2}{\xi(x)} - \eta'(x)\right)y\right\}}{\left\{\frac{dy}{dx} - \frac{\eta(x)}{\xi(x)}y\right\}}, \tag{5.19}$$

and on incorporating the denominator of (5.19) into the arbitrary function of (5.16) we deduce the required second order equation

$$\frac{d^2 y}{dx^2} + \left(\frac{\xi'(x)}{\xi(x)} - 2\frac{\eta(x)}{\xi(x)}\right)\frac{dy}{dx} + \left(\frac{\eta(x)^2}{\xi(x)^2} - \frac{\eta'(x)}{\xi(x)}\right)y = \frac{\Phi_1(A, B)}{s(x)\xi(x)^2}, \tag{5.20}$$

where Φ_1 denotes an arbitrary function and A and B are given by (5.18).

The most general linear homogeneous second order differential equation invariant under the given group is obtained by taking $\Phi_1(A, B)$ to be given by

$$\Phi_1(A, B) = \alpha A + \beta B, \tag{5.21}$$

where α and β are constants. We find from (5.18), (5.20) and (5.21) that this equation becomes

$$\frac{d^2 y}{dx^2} + a(x)\frac{dy}{dx} + b(x)y = 0,$$

where

$$a(x) = \frac{1}{\xi(x)}\{\xi'(x) - 2\eta(x) - \beta\},$$

$$b(x) = \frac{1}{\xi(x)^2}\{\eta(x)^2 - \xi(x)\eta'(x) - \alpha + \beta\eta(x)\},$$

and on eliminating $\eta(x)$ between these two latter equations we obtain

$$\frac{1}{4}(2\xi\xi'' - \xi'^2) + \xi^2\left(b(x) - \frac{1}{2}\frac{da}{dx} - \frac{a(x)^2}{4}\right) = \alpha - \frac{\beta^2}{4}.$$

Hence this result is consistent with that obtained in Section 3.3 (see also Problem 8 part (i) of Chapter 3).

Example 5.6 Find the most general second order differential equation invariant under the group

$$\xi(x,y) = \xi(x)e^{ky}, \quad \eta(x,y) = \eta(x)e^{ky},$$

where k is a constant.

From Example 4.4 we can deduce

$$A = y + \log s(x),$$

$$B = e^{-ky}\left\{\left(\xi(x)\frac{dy}{dx} - \eta(x)\right)^{-1} - \frac{k}{s(x)^k}\int_{x_0}^{x}\frac{s(t)^k}{\xi(t)}dt\right\}, \quad (5.22)$$

with the usual definition for $s(x)$. On differentiating we can show

$$\frac{dB}{dA} = \frac{-\xi(x)^2 e^{-ky}}{(\xi(x)dy/dx - \eta(x))^3}\left\{\frac{d^2y}{dx^2} + k\left(\frac{dy}{dx}\right)^2 + \left(\frac{\xi'(x)}{\xi(x)} - k\frac{\eta(x)}{\xi(x)}\right)\frac{dy}{dx} - \frac{\eta'(x)}{\xi(x)}\right\} - kB,$$

and thus the required differential equation is

$$\frac{d^2y}{dx^2} + k\left(\frac{dy}{dx}\right)^2 + \left(\frac{\xi'(x)}{\xi(x)} - k\frac{\eta(x)}{\xi(x)}\right)\frac{dy}{dx} = \frac{\eta'(x)}{\xi(x)} + \frac{e^{ky}}{\xi(x)^2}\left(\xi(x)\frac{dy}{dx} - \eta(x)\right)^3 \Phi_1(A,B), \quad (5.23)$$

where Φ_1 denotes an arbitrary function and A and B are given by (5.22). Hence a differential equation of the form (5.23) can be reduced to the first order equation

$$\frac{dB}{dA} + kB + \Phi_1(A,B) = 0.$$

Second and Higher Order Ordinary Differential Equations

Example 5.7 Obtain the most general linear third order equation of the form

$$\frac{d^3y}{dx^3} + p(x)\frac{dy}{dx} + q(x)y = 0, \tag{5.24}$$

which is invariant under the group given in Example 5.5.

In the notation of Example 5.5 we have from (5.18) and (5.19)

$$B\frac{dB}{dA} = s(x)\left\{\xi(x)^2\frac{d^2y}{dx^2} + (\xi(x)\xi'(x) - 2\xi(x)\eta(x))\frac{dy}{dx} + (\eta(x)^2 - \xi(x)\eta'(x))y\right\}.$$

On differentiating this equation with respect to A and multiplying the result by B we obtain

$$B\left\{B\frac{d^2B}{dA^2} + \left(\frac{dB}{dA}\right)^2\right\} = s(x)\left\{\xi(x)^3\frac{d^3y}{dx^3} + 3\xi(x)^2(\xi'(x) - \eta(x))\frac{d^2y}{dx^2}\right.$$

$$+ \xi(x)(\xi'(x)^2 + \xi(x)\xi''(x) + 3\eta(x)^2 - 3\eta(x)\xi'(x) - 3\xi(x)\eta'(x))\frac{dy}{dx}$$

$$\left. + (3\xi(x)\eta(x)\eta'(x) - \xi(x)^2\eta''(x) - \xi(x)\xi'(x)\eta'(x) - \eta(x)^3)y\right\}.$$

Hence the most general third order equation invariant under the given group is obtained by equating the expression on the right-hand side of this equation $\Psi_1(A, B, dB/dA)$. The most general linear equation arises from

$$\Psi_1\left(A, B, \frac{dB}{dA}\right) = \alpha A + \beta B + \gamma B\frac{dB}{dA},$$

where α, β and γ are constants. In order to obtain (5.24) we require $\eta(x) = \xi'(x) - \gamma/3$ and we find

$$p(x) = \frac{(\xi'^2 - 2\xi\xi'')}{\xi^2} - \frac{1}{\xi^2}\left(\beta + \frac{\gamma^2}{3}\right),$$

$$q(x) = \frac{(2\xi\xi'\xi'' - \xi^2\xi''' - \xi'^3)}{\xi^3} + \frac{\xi'}{\xi^3}\left(\beta + \frac{\gamma^2}{3}\right) - \frac{1}{\xi^3}\left(\alpha + \frac{\beta\gamma}{3} + \frac{2\gamma^3}{27}\right),$$

and thus

$$\frac{1}{2}\frac{dp}{dx} - q(x) = \frac{1}{\xi(x)^3}\left(\alpha + \frac{\beta\gamma}{3} + \frac{2\gamma^3}{27}\right),$$

which agrees with the result obtained in Section 3.4.

5.5 Applications

In this section we consider three specific second order differential equations which arise from various problems in the nuclear industry. For additional applications the reader should consult Bluman and Cole (1974)(page 116). For our purposes the problems considered illustrate the scope and limitations of the group approach for problems arising from a practical context.

Example 5.8 Reactor core optimization For the problem of determining the appropriate fuel distribution which minimizes the ratio of the critical mass of the core to the reactor power when the power output is prescribed, the following differential equation is obtained

$$y\left(y'' + \frac{\alpha y'}{x}\right) - y'^2 + \beta y^3 = 0, \tag{5.25}$$

where $y(x)$ denotes the non-dimensional thermal flux, primes denote differentiation with respect to x and α and β are known constants. The two cases considered are $\alpha = 0$ which corresponds to assuming a slab geometry while $\alpha = 1$ corresponds to a cylindrical geometry.

For $\alpha = 0$ equation (5.25) becomes

$$yy'' - y'^2 + \beta y^3 = 0, \tag{5.26}$$

and it is instructive to solve this equation first by standard devices and then by the group approach. Since (5.26) does not depend explicitly on x we let $z = y'$ and we obtain in the usual way the first order differential equation

$$\frac{dz}{dy} - \frac{z}{y} = -\frac{\beta y^2}{z}. \tag{5.27}$$

We recognise this as an equation of the Bernoulli type and therefore set $\omega = z^2$ and deduce

$$\omega = C_1 y^2 - 2\beta y^3,$$

where C_1 denotes an integration constant. On integrating

$$\frac{dy}{y(C_1 - 2\beta y)^{1/2}} = dx,$$

we obtain in a straightforward manner

$$y(x) = \frac{C_1}{\beta[1 + \cosh(\sqrt{C_1}x + C_2)]}, \tag{5.28}$$

where C_2 denotes a further integration constant and we are assuming the constant C_1 is positive. Alternatively we may deduce the general solution (5.28) from the group approach in the following way. We observe that (5.26) remains invariant under the two one-parameter groups,

$$x_1 = x + \epsilon, \quad y_1 = y, \tag{5.29}$$

and

$$x_1 = e^\epsilon x, \quad y_1 = e^{-2\epsilon} y. \tag{5.30}$$

The group (5.29) is merely the formal statement that (5.26) does not depend explicitly on x and therefore since y' is an invariant of this group we again set $z = y'$ and obtain (5.27). However the group (5.30) means that (5.27) is invariant under

$$y_1 = e^{-2\epsilon} y, \quad z_1 = e^{-3\epsilon} z,$$

and therefore we select $u^* = z y^{-3/2}$ as the new dependent variable and (5.27) becomes the separable first order differential equation

$$\frac{du^*}{dy} = -\frac{(u^{*2} + 2\beta)}{y u^*}.$$

This equation readily integrates to yield

$$u^{*2} = C_1/y^2 - 2\beta,$$

from which the solution obtained previously can be deduced.

For α non-zero equation (5.25) is still invariant under (5.30) and we therefore select $u = yx^2$ as the new dependent variable. With this substitution, $t = \log x$ and $p = du/dt$ equation (5.25) becomes the Abel equation of the second kind (Murphy (1960), page 25)

$$p\frac{dp}{du} = 2(\alpha - 1)u - \beta u^2 - (\alpha - 1)p + \frac{p^2}{u}.$$

This equation can be reduced to standard from (see equation (1.21)) by the substitution $p = qu$. We find

$$q\frac{dq}{du} = \frac{2(\alpha - 1)}{u} - \beta - (\alpha - 1)\frac{q}{u}, \tag{5.31}$$

which for $\alpha = 1$ can be integrated to finally obtain the following solution for $y(x)$, namely

$$y(x) = \frac{2C_1}{\beta x^2 [2 + x^{\sqrt{C_1}} e^{C_2} + x^{-\sqrt{C_1}} e^{-C_2}]}, \tag{5.32}$$

where again C_1 and C_2 denote integration constants. We note however for other values of α the solution of (5.31) is by no means apparent.

The next two examples arise from the steady state heat conduction equation with non-linear thermal conductivity $k(T)$ and non-linear source $S(T)$, that is

$$\text{div}[k(T)\,\text{grad}\,T] + S(T) = 0, \qquad (5.33)$$

where T denotes the temperature. These problems occur in the context of thermal instability phenomena in rods and plates in the sense that if the rate of energy produced by the heat source exceeds the rate at which energy can be transferred out across the boundary then a steady state temperature distribution cannot exist. The particular thermal conductivity and source terms considered are,

$$k(T) = k_0(T/T_0)^\gamma, \qquad (5.34)$$

$$S(T) = S_0(T/T_0)^\delta, \quad S(T) = S_0\exp(T/T_0), \qquad (5.35)$$

where γ, δ, k_0, S_0 and T_0 denote constants.

Example 5.9 Power law conductivity and source term From (5.33), (5.34) and (5.35)$_1$ the following differential equation in non-dimensional variables may be deduced,

$$y^\gamma\left(y'' + \frac{\alpha y'}{x}\right) + \gamma y'^2 y^{\gamma-1} + \beta y^\delta = 0, \qquad (5.36)$$

where α and β are constants and again $\alpha = 0$ corresponds to the plate or slab geometry while $\alpha = 1$ corresponds to a rod or cylindrical geometry. The three constants α, γ and δ encompass a wide variety of physical behaviour and the full analysis of (5.36) involves consideration of a number of special cases. Here we restrict our attention to results which can be deduced rapidly and simply from the group approach. For any practical problem we first examine simple groups leaving the equation invariant before using the theory given in the first section of this chapter and equation (5.8). For the examples of this section, this aspect is summarized in Problems 9, 10 and 11.

If we look for a simple stretching group

$$x_1 = e^\epsilon x, \quad y_1 = e^{a\epsilon} y, \qquad (5.37)$$

leaving (5.36) invariant then we find

$$a = \frac{2}{(1+\gamma-\delta)}, \qquad (5.38)$$

provided $\delta \neq 1 + \gamma$. Assuming for the time being that this is the case we take $u = yx^{-a}$ as the new dependent variable so that equation (5.36) becomes

$$u^\gamma\{x^2 u'' + (\alpha + 2a)ux' + a(\alpha + a - 1)u\} + \gamma u^{\gamma-1}(xu' + au)^2 + \beta u^\delta = 0,$$

and with the substitutions $t = \log x$ and $p = du/dt$ this equation becomes

$$up\frac{dp}{du} + (\alpha + 2a + 2a\gamma - 1)pu + a(\alpha + a + \gamma a - 1)u^2 + \gamma p^2 + \beta u^{-2/a} = 0. \tag{5.39}$$

In general this is again an Abel equation of the second kind. However special cases give rise to standard equations. For example if $\delta = \gamma + 3$ the equation is homogeneous while if $\alpha \neq 1$ and $\delta = (1+\gamma)(\alpha+3)/(\alpha-1)$ the equation is of the Bernoulli type.

If $\delta = 1 + \gamma$ then (5.36) becomes

$$y\left(y'' + \frac{\alpha y'}{x}\right) + \gamma y'^2 + \beta y^2 = 0, \tag{5.40}$$

which is invariant under the one-parameter group

$$x_1 = x, \quad y_1 = e^\epsilon y. \tag{5.41}$$

Thus we take $w = y'/y$ as the new dependent variable (see Example 5.4) and (5.40) becomes

$$\frac{dw}{dx} + \frac{\alpha}{x}w + (1+\gamma)w^2 + \beta = 0, \tag{5.42}$$

which is a Ricatti equation (Murphy (1960), page 15). Clearly $\alpha = 0$ or $\gamma = -1$ are special cases which can be readily solved. The general solution† of (5.42) or (5.40) may be deduced as follows. Introduce the variable $Y = y^{1+\gamma}$ or $y = Y^{(1+\gamma)^{-1}}$, then equation (5.40) becomes

$$Y'' + \frac{\alpha}{x}Y' + \beta(1+\gamma)Y = 0.$$

Further by means of the transformation

$$Y(x) = x^{(1-\alpha)/2} Z(x),$$

this equation becomes

$$Z'' + \frac{Z'}{x} + \left\{\beta(1+\gamma) - \frac{(\alpha-1)^2}{4x^2}\right\}Z = 0,$$

---

† The author wishes to express his gratitude to D.K. Kalra (Indian Institute of Technology, New Delhi) for pointing out that (5.42) admits this general solution.

which can be recognised as a Bessel equation with solution

$$Z(x) = C_1 J_\lambda \left[x\sqrt{\beta(1+\gamma)}\right] + C_2 Y_\lambda \left[x\sqrt{\beta(1+\gamma)}\right],$$

where $\lambda = (\alpha - 1)/2$ and assuming $\beta(1+\gamma)$ is positive. Thus if $\delta = 1+\gamma$ a closed form solution is possible for all values of α, β and γ.

For the final example we consider the equation arising from (5.33) for the case of constant thermal conductivity and exponential source term.

Example 5.10 Constant conductivity and exponential source term From (5.33), (5.34) ($\gamma = 0$) and (5.35)$_2$ and with non-dimensional variables we may deduce the equation

$$y'' + \frac{\alpha y'}{x} = \beta e^y. \tag{5.43}$$

For α zero this equation can be readily integrated by means of the standard substitution $z = dy/dx$. For α non-zero we look for a one-parameter group leaving (5.43) invariant of the form

$$x_1 = e^\epsilon x, \quad y_1 = y + a\epsilon. \tag{5.44}$$

We find $a = -2$ and therefore on eliminating ϵ from (5.44) we obtain $u = x^2 e^y$ as an invariant of the group. With u as the dependent variable (5.43) becomes

$$x^2(uu'' - u'^2) + \alpha x u u' + 2(1-\alpha)u^2 = \beta u^3,$$

and the usual substitutions $t = \log x$ and $p = du/dt$ yield,

$$up\frac{dp}{du} = p^2 + (1-\alpha)up - 2(1-\alpha)u^2 + \beta u^3. \tag{5.45}$$

For $\alpha = 1$ equation (5.45) is the same as equation (5.27) and therefore the solution can be deduced from (5.28). However for $\alpha \neq 1$ (5.45) must be solved as an Abel equation of the second kind.

The examples of this section illustrate how simple groups leaving the equation invariant may be utilized to reduce the order of the differential equation. The resulting differential equations may or may not be of a standard type with a simple solution.

PROBLEMS

1. In the notation of Example 5.1 show that if $\rho(x)$ is non-zero then

$$F(x,y) = G(x)y + H(x),$$

where

$$G(x) = \frac{\rho''(x)}{\rho(x)}, \quad H(x) = \frac{[2\eta'(x) - \xi''(x)]}{3\rho(x)},$$

and deduce that $\rho(x)$, $\xi(x)$ and $\eta(x)$ must be such that

$$\eta(x) + \xi'(x) - 3\frac{\rho'(x)}{\rho(x)}\xi(x) = C_1 \int_{x_0}^{x} \frac{dt}{\rho(t)^2} + C_2,$$

for constants C_1 and C_2.

2. **Continuation.** If $\rho(x)$ is identically zero and $2\eta(x) = \xi'(x)$ show that $F(x,y)$ is given by

$$F(x,y) = \frac{(\xi(x)^{1/2})''}{\xi(x)^{1/2}} y + \frac{1}{\xi(x)^{1/2}} \left(\frac{\zeta(x)}{\xi(x)^{1/2}}\right)' + \frac{1}{\xi(x)^{3/2}} \Phi\left(\frac{y}{\xi(x)^{1/2}} - \int_{x_0}^{x} \frac{\zeta(t)}{\xi(t)^{3/2}} dt\right),$$

where Φ denotes an arbitrary function of the argument indicated.

3. If $p(x)$ is non-zero but arbitrary show that the differential equation

$$\frac{d^2y}{dx^2} + p(x)y^2 = 0,$$

is invariant under at most six one-parameter groups. Show that

$$\frac{d^2y}{dx^2} + \alpha x^m y^2 = 0,$$

can be reduced to a first order Abel equation of the second kind (Murphy (1960), page 25).

4. Find the most general second order differential equations which are invariant under the one-parameter groups given in Problems 5 and 7 of Chapter 4.

5. Show that the second order differential equation

$$F(x, y, y', y'') = 0,$$

is invariant under the one-parameter group

$$x_1 = x + \epsilon \xi(x,y) + O(\epsilon^2), \quad y_1 = y + \epsilon \eta(x,y) + O(\epsilon^2), \qquad (*)$$

if and only if

$$L''F = 0,$$

where L'' is the second extension of the operator L, namely

$$L'' = \xi \frac{\partial}{\partial x} + \eta \frac{\partial}{\partial y} + \pi \frac{\partial}{\partial y'} + \sigma \frac{\partial}{\partial y''},$$

where π and σ are the infinitesimal versions of y' and y'' respectively and are defined by (5.5) and (5.6). (Note that if $A(x,y)$, $B(x,y,y')$ and $C(x,y,y',y'')$ are three independent integrals of the equations

$$\frac{dx}{d\tau} = \xi(x,y), \quad \frac{dy}{d\tau} = \eta(x,y), \quad \frac{dy'}{d\tau} = \pi(x,y,y'), \quad \frac{dy''}{d\tau} = \sigma(x,y,y',y''),$$

then the most general second order equation invariant under $(*)$ takes the form $\Omega(A,B,C) = 0$ or $C = \Phi(A,B)$.)

[For a detailed discussion of the following three problems the reader should consult Dickson (1924), page 358.]

6. Given two one-parameter groups with operators

$$L_1 = \xi_1(x,y)\frac{\partial}{\partial x} + \eta_1(x,y)\frac{\partial}{\partial y}, \quad L_2 = \xi_2(x,y)\frac{\partial}{\partial x} + \eta_2(x,y)\frac{\partial}{\partial y},$$

show that the first extension of the commutator $(L_1 L_2)'$ is identical to the commutator of their respective first extensions, that is $(L_1' L_2')$. (See Problems 11 and 12 of Chapter 4.)

7. **Continuation.** Show that if the second order differential equation, $y'' = F(x,y,y')$ is invariant under two one-parameter groups with operators L_1 and L_2 then it is also invariant under the one-parameter group with operator $(L_1 L_2)$.

8. **Continuation.** If L_1 and L_2 leave $y'' = F(x,y,y')$ invariant show that there exists an operator L_3 which also leaves the equation invariant and is such that

$$(L_1 L_3) = aL_1 + bL_3,$$

for some constants a and b.

Second and Higher Order Ordinary Differential Equations

9. For the second order differential equation

$$\frac{d^2y}{dx^2} = -\frac{\alpha}{x}\frac{dy}{dx} - \frac{\gamma}{y}\left(\frac{dy}{dx}\right)^2 - f(y), \qquad (+)$$

where α and γ are constants such that $\gamma \neq -1$, show that for all functions $f(y)$ the only one-parameter groups leaving $(+)$ invariant take the form

$$\xi(x,y) = g(x),$$
$$\eta(x,y) = \left\{\frac{1}{2}\left[g'(x) - \frac{\alpha}{x}g(x)\right] + A\right\}\frac{y}{(1+\gamma)} + \frac{h(x)}{y^\gamma}, \qquad (++)$$

where A denotes an arbitrary constant and $g(x)$ and $h(x)$ are functions of x such that

$$\frac{\partial^2 \eta}{\partial x^2} + \frac{\alpha}{x}\frac{\partial \eta}{\partial x} = \left\{\frac{\partial \eta}{\partial y} - 2g'(x)\right\}f(y) - \eta f'(y).$$

10. **Continuation.** If $f(y) = \beta y^{\delta-\gamma}$ show that if $\delta \neq \gamma + 1$ the only one-parameter group leaving $(+)$ invariant is given by

$$\xi(x,y) = x, \qquad \eta(x,y) = \frac{2y}{(1+\gamma-\delta)}.$$

If $\delta = \gamma + 1$ show that the differential equation $(+)$ is invariant under the group $(++)$ where $g(x)$ and $h(x)$ satisfy the following differential equations

$$g''' + 4\beta(1+\gamma)g' + \alpha(2-\alpha)\left(\frac{g'}{x^2} - \frac{g}{x^3}\right) = 0,$$

$$h'' + \frac{\alpha}{x}h' + \beta(1+\gamma)h = 0.$$

11. **Continuation.** If $f(y) = \beta e^y$ and $\gamma = 0$ show that for $\alpha \neq 1$ the only one-parameter group leaving $(+)$ invariant is given by $(++)$ with $g(x)$ and $h(x)$ given by

$$g(x) = \frac{2A}{(\alpha-1)}x, \qquad h(x) = -\frac{4A}{(\alpha-1)},$$

where A is the arbitrary constant in Problem 9. For $\alpha = 1$ show that $g(x)$ and $h(x)$ are given by

$$g(x) = -2Ax\log x + Bx, \qquad h(x) = 4A\log x + 2(2A - B),$$

where B is a further arbitrary constant.

12. For the differential equation (5.25) of Example 5.8, show that
    (i) $y = 2(\alpha - 1)/\beta x^2$ is a special solution,
    (ii) the transformation $y = e^Z$ yields the following differential equation
    $$Z'' + \frac{\alpha}{x}Z' + \beta e^Z = 0.$$

13. For the differential equation (5.36) of Example 5.9 show that,
    (i) $y = Ax^B$ provides a special solution where A and B are determined from
    $$B = \frac{2}{(1+\gamma-\delta)}, \quad A = \left(\frac{\beta}{B[1-\alpha-(1+\gamma)B]}\right)^{1/(1+\gamma-\delta)}$$
    (ii) the transformation $y = Z^{1/(1+\gamma)}$ yields the following differential equation
    $$Z'' + \frac{\alpha}{x}Z' + \beta(1+\gamma)Z^{\delta/(1+\gamma)} = 0.$$

    [Notice that if $\alpha = 2$, then this equation is known as the Lane-Emden equation of index $\delta/(1+\gamma)$. Special values of this index give rise to integrable cases, see Murphy (1960), page 387.]

14. Show that the classical diffusion equation
    $$\frac{\partial c}{\partial t} = D\frac{\partial^2 c}{\partial x^2},$$
    admits travelling wave solutions of the form
    $$c(x,t) = A(x)\sin[\omega t - B(x)],$$
    where ω is a constant and $A(x)$ and $B(x)$ satisfy
    $$A'' = AB'^2, \quad AB'' + 2A'B' + k^2 A = 0,$$
    where $k = (\omega/D)^{1/2}$. Observe that
    $$[(A^3 A'')^{1/2}]' + k^2 A^2 = 0,$$
    remains invariant under the one-parameter group
    $$x_1 = x, \quad A_1 = e^\epsilon A,$$

Second and Higher Order Ordinary Differential Equations   97

and deduce the second order differential equation

$$w'' + 6ww' + 4w^3 + 2k^2(w' + w^2)^{1/2} = 0,$$

where $w = A'/A$.

15. **Continuation.** Introduce $\Theta(x)$ and the complex variable $Z(x)$ by

$$A(x) = \exp \Theta(x), \quad Z(x) = \Theta(x) + iB(x),$$

and show that the coupled equations for $A(x)$ and $B(x)$ give

$$Z'' + Z'^2 = -ik^2.$$

Next introduce $p = Z'$ and $q = p^2$ and deduce the standard linear equation,

$$\frac{dq}{dz} + 2q = -2ik^2,$$

so that

$$q = Ce^{-2Z} - ik^2,$$

where C denotes a complex arbitrary constant. From this equation with

$$C = ik^2 A_0^2 e^{2iB_0},$$

where A_0 and B_0 denote arbitrary real constants, deduce that $z = e^Z$ is given by

$$z = A_0 e^{iB_0} \sin[k(1+i)(x-x_0)/2^{1/2}],$$

where x_0 is a further real integration constant. Hence conclude that the general solutions for $A(x)$ and $B(x)$ are given by

$$A(x) = A_0(\sin^2 y + \sinh^2 y)^{1/2},$$

$$B(x) = B_0 + \tan^{-1}\left(\frac{\tanh y}{\tan y}\right),$$

where y denotes $k(x - x_0)/2^{1/2}$. Show that $c(x,t)$ simplifies to give

$$c(x,t) = \frac{A_0}{2}\{e^y \cos(\omega t - B_0 - y) - e^{-y}\cos(\omega t - B_0 + y)\}.$$

Chapter Six
Linear partial differential equations

6.1 Introduction

For partial differential equations the calculations involved in the determination of a one-parameter group leaving the equation invariant are generally fairly lengthy. In order to keep these calculations to a minimum we first consider a restricted class of one-parameter transformation groups applicable to linear partial differential equations. Non-linear equations are considered in the following chapter. Specifically for a single dependent variable c and two independent variables x and t we consider transformations of the form

$$\begin{aligned} x_1 &= f(x,t,\epsilon) = x + \epsilon\xi(x,t) + \mathbf{O}(\epsilon^2), \\ t_1 &= g(x,t,\epsilon) = t + \epsilon\eta(x,t) + \mathbf{O}(\epsilon^2), \\ c_1 &= h(x,t,\epsilon)c = c + \epsilon\zeta(x,t)c + \mathbf{O}(\epsilon^2), \end{aligned} \quad (6.1)$$

where the functions f, g and h do not depend explicitly on c. If the transformation (6.1) leaves a given partial differential equation invariant and if $c = \phi(x,t)$ then from $c_1 = \phi(x_1,t_1)$ on equating terms of order ϵ we have

$$\xi(x,t)\frac{\partial c}{\partial x} + \eta(x,t)\frac{\partial c}{\partial t} = \zeta(x,t)c. \quad (6.2)$$

For known functions $\xi(x,t)$, $\eta(x,t)$ and $\zeta(x,t)$, equation (6.2) when solved as a first order partial differential equation, yields the functional form of the similarity solution in terms of an arbitrary function. This arbitrary function is determined by substitution of the functional form of the solution into the given partial differential equation. In the case of two independent variables the resulting equation is an ordinary differential equation. For more than two independent variables the procedure reduces the number of indpendent variables by one.

In the following section we give the formulae for the infinitesimal versions of the partial derivatives $\partial c/\partial x$, $\partial c/\partial t$, $\partial^2 c/\partial x^2$, $\partial^2 c/\partial x \partial t$ and $\partial^2 c/\partial t^2$. Although we make no use of the last two partial derivatives, they are included for completeness. For the remainder of the chapter we principally consider groups of the form (6.1) and the corresponding solutions of diffusion equations. In particular we consider the classical diffusion equation

$$\frac{\partial c}{\partial t} = \frac{\partial^2 c}{\partial x^2}, \quad (6.3)$$

and the Fokker-Planck equation which we assume given in the form,

$$\frac{\partial c}{\partial t} = \frac{\partial}{\partial x}\left(p(x)\frac{\partial c}{\partial x}\right) + \frac{\partial}{\partial x}(q(x)c), \qquad (6.4)$$

where $p(x)$ and $q(x)$ are functions of x only. In the determination of groups leaving an equation invariant there are two methods, termed *classical* and *non-classical*. The classical approach equates the infinitesimal version of the given partial differential equation to zero without making use of equation (6.2). The non-classical procedure which is considerably more complicated makes use of (6.2) and includes the classical groups as special cases. For the most part we obtain results from the classical procedure. However in the final section we discuss the non-classical approach with reference to equation (6.3). The results given in this chapter for (6.3) can also be found in Bluman and Cole (1974)(page 206). In Section 6.5 we present some results for equation (6.4). We show for arbitrary $p(x)$ how the most general function $q(x)$ can be found such that (6.4) admits a classical group of transformations leaving the equation invariant.

6.2 Formulae for partial derivatives

For the one-parameter group of transformations (6.1) we assume that the Jacobian,

$$J = \frac{\partial(x_1, t_1)}{\partial(x, t)} = \frac{\partial x_1}{\partial x}\frac{\partial t_1}{\partial t} - \frac{\partial x_1}{\partial t}\frac{\partial t_1}{\partial x}, \qquad (6.5)$$

is non-zero and finite. From (6.1) and (6.5) we have

$$J = 1 + \epsilon\left(\frac{\partial \xi}{\partial x} + \frac{\partial \eta}{\partial t}\right) + \mathbf{O}(\epsilon^2). \qquad (6.6)$$

Now for the partial derivative $\partial c_1/\partial x_1$ we have

$$\frac{\partial c_1}{\partial x_1} = \frac{\partial(c_1, t_1)}{\partial(x_1, t_1)} = \frac{1}{J}\frac{\partial(c_1, t_1)}{\partial(x, t)}, \qquad (6.7)$$

and on substituting $(6.1)_2$, $(6.1)_3$ and (6.6) into (6.7) we obtain

$$\frac{\partial c_1}{\partial x_1} = \frac{\partial c}{\partial x} + \epsilon\left\{\frac{\partial(\zeta c)}{\partial x} + \frac{\partial(c, \eta)}{\partial(x, t)} - \left(\frac{\partial \xi}{\partial x} + \frac{\partial \eta}{\partial t}\right)\frac{\partial c}{\partial x}\right\} + \mathbf{O}(\epsilon^2),$$

which simplifies to give

$$\frac{\partial c_1}{\partial x_1} = \frac{\partial c}{\partial x} + \epsilon\left\{c\frac{\partial \zeta}{\partial x} + \left(\zeta - \frac{\partial \xi}{\partial x}\right)\frac{\partial c}{\partial x} - \frac{\partial \eta}{\partial x}\frac{\partial c}{\partial t}\right\} + \mathbf{O}(\epsilon^2). \qquad (6.8)$$

Linear Partial Differential Equations

Similarly from

$$\frac{\partial c_1}{\partial t_1} = -\frac{\partial(c_1, x_1)}{\partial(x_1, t_1)} = -\frac{1}{J}\frac{\partial(c_1, x_1)}{\partial(x, t)}, \tag{6.9}$$

we obtain

$$\frac{\partial c_1}{\partial t_1} = \frac{\partial c}{\partial t} + \epsilon \left\{ c\frac{\partial \zeta}{\partial t} + \left(\zeta - \frac{\partial \eta}{\partial t}\right)\frac{\partial c}{\partial t} - \frac{\partial \xi}{\partial t}\frac{\partial c}{\partial x} \right\} + \mathbf{O}(\epsilon^2). \tag{6.10}$$

If we introduce π_1 and π_2 by

$$\begin{aligned}\pi_1 &= \left\{ c\frac{\partial \zeta}{\partial x} + \left(\zeta - \frac{\partial \xi}{\partial x}\right)\frac{\partial c}{\partial x} - \frac{\partial \eta}{\partial x}\frac{\partial c}{\partial t} \right\}, \\ \pi_2 &= \left\{ c\frac{\partial \zeta}{\partial t} + \left(\zeta - \frac{\partial \eta}{\partial t}\right)\frac{\partial c}{\partial t} - \frac{\partial \xi}{\partial t}\frac{\partial c}{\partial x} \right\}, \end{aligned} \tag{6.11}$$

then (6.8) and (6.10) become respectively

$$\frac{\partial c_1}{\partial x_1} = \frac{\partial c}{\partial x} + \epsilon\pi_1 + \mathbf{O}(\epsilon^2), \quad \frac{\partial c_1}{\partial t_1} = \frac{\partial c}{\partial t} + \epsilon\pi_2 + \mathbf{O}(\epsilon^2). \tag{6.12}$$

For the second order partial derivative $\partial^2 c_1/\partial x_1^2$ we have

$$\frac{\partial^2 c_1}{\partial x_1^2} = \frac{\partial\left(\frac{\partial c_1}{\partial x_1}, t_1\right)}{\partial(x_1, t_1)} = \frac{1}{J}\frac{\partial\left(\frac{\partial c_1}{\partial x_1}, t_1\right)}{\partial(x, t)}.$$

From (6.6) and (6.12)$_1$ we deduce

$$\frac{\partial^2 c_1}{\partial x_1^2} = \frac{\partial^2 c}{\partial x^2} + \epsilon\left\{ \frac{\partial \pi_1}{\partial x} + \frac{\partial(\partial c/\partial x, \eta)}{\partial(x,t)} - \left(\frac{\partial \xi}{\partial x} + \frac{\partial \eta}{\partial t}\right)\frac{\partial^2 c}{\partial x^2} \right\} + \mathbf{O}(\epsilon^2),$$

which on using (6.11)$_1$ becomes

$$\begin{aligned}\frac{\partial^2 c_1}{\partial x_1^2} = \frac{\partial^2 c}{\partial x^2} + \epsilon \Bigg\{ & c\frac{\partial^2 \zeta}{\partial x^2} + \left(2\frac{\partial \zeta}{\partial x} - \frac{\partial^2 \xi}{\partial x^2}\right)\frac{\partial c}{\partial x} - \frac{\partial^2 \eta}{\partial x^2}\frac{\partial c}{\partial t} \\ & + \left(\zeta - 2\frac{\partial \xi}{\partial x}\right)\frac{\partial^2 c}{\partial x^2} - 2\frac{\partial \eta}{\partial x}\frac{\partial^2 c}{\partial x \partial t} \Bigg\} + \mathbf{O}(\epsilon^2). \end{aligned} \tag{6.13}$$

Similarly for $\partial^2 c_1/\partial x_1 \partial t_1$ and $\partial^2 c_1/\partial t_1^2$ we can deduce from the equations

$$\frac{\partial^2 c_1}{\partial x_1 \partial t_1} = \frac{\partial\left(\frac{\partial c_1}{\partial t_1}, t_1\right)}{\partial(x_1, t_1)} = \frac{1}{J}\frac{\partial\left(\frac{\partial c_1}{\partial t_1}, t_1\right)}{\partial(x, t)},$$

$$\frac{\partial^2 c_1}{\partial t_1^2} = -\frac{\partial\left(\frac{\partial c_1}{\partial t_1}, x_1\right)}{\partial(x_1, t_1)} = -\frac{1}{J}\frac{\partial\left(\frac{\partial c_1}{\partial t_1}, x_1\right)}{\partial(x, t)},$$

the following results,

$$\frac{\partial^2 c_1}{\partial x_1 \partial t_1} = \frac{\partial^2 c}{\partial x \partial t} + \epsilon \left\{ c \frac{\partial^2 \zeta}{\partial x \partial t} + \left(\frac{\partial \zeta}{\partial t} - \frac{\partial^2 \xi}{\partial x \partial t} \right) \frac{\partial c}{\partial x} + \left(\frac{\partial \zeta}{\partial x} - \frac{\partial^2 \eta}{\partial t \partial x} \right) \frac{\partial c}{\partial t} \right.$$
$$\left. - \frac{\partial \xi}{\partial t} \frac{\partial^2 c}{\partial x^2} + \left(\zeta - \frac{\partial \xi}{\partial x} - \frac{\partial \eta}{\partial t} \right) \frac{\partial^2 c}{\partial x \partial t} - \frac{\partial \eta}{\partial x} \frac{\partial^2 c}{\partial t^2} \right\} + \mathbf{O}(\epsilon^2),$$

$$\frac{\partial^2 c_1}{\partial t_1^2} = \frac{\partial^2 c}{\partial t^2} + \epsilon \left\{ c \frac{\partial^2 \zeta}{\partial t^2} - \frac{\partial^2 \xi}{\partial t^2} \frac{\partial c}{\partial x} + \left(2 \frac{\partial \zeta}{\partial t} - \frac{\partial^2 \eta}{\partial t^2} \right) \frac{\partial c}{\partial t} \right.$$
$$\left. - 2 \frac{\partial \xi}{\partial t} \frac{\partial^2 c}{\partial x \partial t} + \left(\zeta - 2 \frac{\partial \eta}{\partial t} \right) \frac{\partial^2 c}{\partial t^2} \right\} + \mathbf{O}(\epsilon^2).$$

6.3 Classical groups for the diffusion equation

In this section we deduce the classical group of the diffusion equation (6.3). From (6.10) and (6.13) we have

$$\frac{\partial c_1}{\partial t_1} - \frac{\partial^2 c_1}{\partial x_1^2} = \frac{\partial c}{\partial t} - \frac{\partial^2 c}{\partial x^2} + \epsilon \left\{ c \left(\frac{\partial \zeta}{\partial t} - \frac{\partial^2 \zeta}{\partial x^2} \right) + \left(\zeta - \frac{\partial \eta}{\partial t} + \frac{\partial^2 \eta}{\partial x^2} \right) \frac{\partial c}{\partial t} \right.$$
$$\left. + \left(-2 \frac{\partial \zeta}{\partial x} - \frac{\partial \xi}{\partial t} + \frac{\partial^2 \xi}{\partial x^2} \right) \frac{\partial c}{\partial x} + \left(2 \frac{\partial \xi}{\partial x} - \zeta \right) \frac{\partial^2 c}{\partial x^2} + 2 \frac{\partial \eta}{\partial x} \frac{\partial^2 c}{\partial x \partial t} \right\} + \mathbf{O}(\epsilon^2). \quad (6.14)$$

If we now make use of (6.3) we find that the diffusion equation remains invariant under the transformation (6.1) provided the functions $\xi(x,t)$, $\eta(x,t)$ and $\zeta(x,t)$ are such that the equation

$$c \left(\frac{\partial \zeta}{\partial t} - \frac{\partial^2 \zeta}{\partial x^2} \right) + \left(2 \frac{\partial \xi}{\partial x} - \frac{\partial \eta}{\partial t} + \frac{\partial^2 \eta}{\partial x^2} \right) \frac{\partial c}{\partial t}$$
$$+ \left(-2 \frac{\partial \zeta}{\partial x} - \frac{\partial \xi}{\partial t} + \frac{\partial^2 \xi}{\partial x^2} \right) \frac{\partial c}{\partial x} + 2 \frac{\partial \eta}{\partial x} \frac{\partial^2 c}{\partial x \partial t} = 0, \quad (6.15)$$

is satisfied identically. For the non-classical approach we simplify (6.15) further by means of (6.2). This is done in the final section of this chapter. For the classical group we simply equate the coefficients of c and its derivatives to zero.

From the coefficient of $\partial^2 c/\partial x \partial t$ we deduce that $\eta = \eta(t)$ while from the coefficient of $\partial c/\partial t$ we have

$$\xi = \frac{\eta'(t)x}{2} + \rho(t), \quad (6.16)$$

Linear Partial Differential Equations

where the prime here denotes differentiation with respect to t and $\rho(t)$ denotes an arbitrary function of t. On equating the coefficient of $\partial c/\partial x$ to zero and making use of (6.16) we deduce that

$$\zeta = -\frac{1}{2}\left(\frac{\eta''(t)x^2}{4} + \rho'(t)x\right) + \sigma(t), \tag{6.17}$$

where $\sigma(t)$ denotes a further arbitrary function of t. From the coefficient of c in equation (6.15) and using (6.17) we obtain

$$-\frac{1}{2}\left(\frac{\eta'''(t)x^2}{4} + \rho''(t)x\right) + \sigma'(t) + \frac{\eta''(t)}{4} = 0,$$

from which it is apparent that we require

$$\eta'''(t) = 0, \quad \rho''(t) = 0, \quad \sigma'(t) = -\frac{\eta''(t)}{4}.$$

From these equations it is now a simple matter to deduce the classical group of the diffusion equation, namely

$$\xi(x,t) = \kappa + \delta t + \beta x + \gamma x t,$$
$$\eta(x,t) = \alpha + 2\beta t + \gamma t^2, \tag{6.18}$$
$$\zeta(x,t) = -\gamma\left(\frac{x^2}{4} + \frac{t}{2}\right) - \frac{\delta x}{2} + \lambda,$$

where α, β, γ, δ, λ and κ denote six arbitrary constants and for comparison purposes we have adopted the same notation used in Bluman and Cole (1974). Some of these constants give rise to standard or even trivial solutions of (6.3). However it is instructive for the reader to deduce the global form of the one-parameter group and the resulting similarity solutions of the diffusion equation. The constants κ, α and λ represent respectively the invariance of (6.3) under translations of x and t and stretching of c (see Problem 1). The constants β, γ and δ are considered in the examples of the following section.

6.4 Simple examples for the diffusion equation

The general classical similarity solution of (6.3) is obtained from (6.2) and (6.18) with all the constants in (6.18) non-zero. For purposes of illustration it is useful to consider the solutions arising from one non-zero constant with the others taken to be zero.

Example 6.1 $\beta = 1$, $\alpha = \gamma = \delta = \lambda = \kappa = 0$. In this case the global form of the one-parameter group is obtained by solving

$$\frac{dx_1}{d\epsilon} = x_1, \quad \frac{dt_1}{d\epsilon} = 2t_1, \quad \frac{dc_1}{d\epsilon} = 0,$$

subject to the initial conditions

$$x_1 = x, \quad t_1 = t, \quad c_1 = c, \qquad (6.19)$$

when $\epsilon = 0$. In this case we find

$$x_1 = e^\epsilon x, \quad t_1 = e^{2\epsilon} t, \quad c_1 = c,$$

so that clearly the constant β reflects the invariance of (6.3) under simultaneous stretchings of x and t. From (6.2) we obtain the partial differential equation

$$x \frac{\partial c}{\partial x} + 2t \frac{\partial c}{\partial t} = 0,$$

which on solving gives rise to the functional form previously considered in Problem 7 of Chapter 1.

Example 6.2 $\delta = 1$, $\alpha = \beta = \gamma = \lambda = \kappa = 0$. In order to deduce the global form of this group we require to solve

$$\frac{dx_1}{d\epsilon} = t_1, \quad \frac{dt_1}{d\epsilon} = 0, \quad \frac{dc_1}{d\epsilon} = -\frac{x_1}{2} c_1,$$

with initial conditions (6.19). We find

$$x_1 = x + \epsilon t, \quad t_1 = t, \quad c_1 = c \exp\left(-\frac{\epsilon x}{2} - \frac{\epsilon^2 t}{4}\right).$$

Further from (6.2) the functional form of the solution is obtained by solving

$$t \frac{\partial c}{\partial x} = -\frac{x}{2} c,$$

which yields

$$c(x, t) = e^{-x^2/4t} \phi(t), \qquad (6.20)$$

where $\phi(t)$ denotes an arbitrary function of t. On substituting (6.20) into (6.3) we readily deduce the ordinary differential equation

$$\phi'(t) + \frac{\phi(t)}{2t} = 0,$$

Linear Partial Differential Equations

and therefore

$$\phi(t) = \frac{\phi_0}{\sqrt{t}},$$

where ϕ_0 denotes an arbitrary constant. From this equation and (6.20) we see that the constant δ also gives rise to the well known source solution determined in Example 1.2.

Example 6.3 $\gamma = 1$, $\alpha = \beta = \delta = \lambda = \kappa = 0$. In this case we have

$$\frac{dx_1}{d\epsilon} = x_1 t_1, \quad \frac{dt_1}{d\epsilon} = t_1^2, \quad \frac{dc_1}{d\epsilon} = -\left(\frac{x_1^2}{4} + \frac{t_1}{2}\right) c_1, \qquad (6.21)$$

together with the initial conditions (6.19). From $(6.21)_2$ we have

$$t_1 = \frac{t}{(1 - \epsilon t)}, \qquad (6.22)$$

and therefore $(6.21)_1$ becomes

$$\frac{dx_1}{d\epsilon} = \frac{x_1 t}{(1 - \epsilon t)},$$

which on integration yields

$$x_1 = \frac{x}{(1 - \epsilon t)}. \qquad (6.23)$$

Using (6.22) and (6.23) in $(6.21)_3$ and integrating the resulting equation we find

$$c_1 = c(1 - \epsilon t)^{1/2} \exp\left(\frac{-\epsilon x^2}{4(1 - \epsilon t)}\right). \qquad (6.24)$$

In order to determine the functional form of the corresponding similarity solution we have from (6.2)

$$xt \frac{\partial c}{\partial x} + t^2 \frac{\partial c}{\partial t} = -\left(\frac{x^2}{4} + \frac{t}{2}\right) c. \qquad (6.25)$$

On solving this equation we find that

$$c(x, t) = \frac{e^{-x^2/4t}}{t^{1/2}} \phi\left(\frac{x}{t}\right), \qquad (6.26)$$

where ϕ denotes an arbitrary function of the argument indicated. On substitution of (6.26) in (6.3) we find that we have simply

$$\phi''\left(\frac{x}{t}\right) = 0,$$

so the constant γ gives rise to the solution

$$c(x,t) = \frac{e^{-x^2/4t}}{t^{1/2}}\left(\phi_0 + \phi_1 \frac{x}{t}\right),$$

which again includes the source solution (1.29) as well as the solution of (6.3) which is the derivative of the source solution with respect to x. Thus although no new solutions are obtained by considering separately the constants in (6.18), these simple examples illustrate the basic procedure in simple terms. In order to obtain non-trivial results we need to consider the full group (6.18). This is done in the following section with reference to moving boundary problems (see also Problems 6, 7, 8 and 9).

6.5 Moving boundary problems

Problems involving the classical diffusion equation (6.3) and an unknown moving boundary $x = X(t)$ occur in many important areas of science, engineering and industry (see for example Hill (1987)). The literature on these problems is scattered throughout many diverse disciplines and it is not possible here to consider the subject in detail. The purpose of this section is to identify the moving boundaries $x = X(t)$ which remain invariant under the classical group (6.18). These boundaries relate to most of the exact analytic results which are available for such problems and therefore might provide a useful guide to the solution of other problems with unknown boundaries.

Typically a moving boundary problem takes the form

$$\frac{\partial c}{\partial t} = \frac{\partial^2 c}{\partial x^2}, \quad 0 < x < X(t),$$

$$c(X(t),t) = 0, \quad \frac{\partial c}{\partial x}(X(t),t) = -\dot{X}(t),$$

(6.27)

together with either c or $\partial c/\partial x$ (or linear combination of these) prescribed on $x = 0$ and prescribed initial data for c and X. We note that the dot denotes diferentiation with respect to time and that for this problem the initial condition is that c is zero. Such problems are non-linear and it is instructive to observe the precise nature of the non-linearity for those problems which can be transformed to fixed boundary value problems. Assuming that both $X(t)$ and $\dot{X}(t)$ are never zero and that the prescribed data for c or $\partial c/\partial x$ on $x = 0$ does not explicitly involve t then we can make the transformation.

$$\rho = \frac{x}{X(t)}, \quad \tau = X(t),$$

(6.28)

Linear Partial Differential Equations

and with $c(x,t) = C(\rho, \tau)$ the moving boundary problem (6.27) becomes

$$\frac{\partial^2 C}{\partial \rho^2} = \frac{\partial C}{\partial \rho}(1,\tau)\left\{\rho\frac{\partial C}{\partial \rho} - \tau\frac{\partial C}{\partial \tau}\right\}, \quad 0 < \rho < 1, \tag{6.29}$$

$$C(1,\tau) = 0, \quad \frac{\partial C}{\partial \rho}(1,\tau) = -\tau\dot{\tau}.$$

Thus in principle we have transformed (6.27) to a fixed boundary value problem except that now the equation to be solved is non-linear, although not of the usual type of non-linearity with which we are familiar. For example, for prescribed data $c(0,t) = c_0$ and $X(0) = a$ where c_0 and a are constants and zero initial condition we supplement (6.29) with

$$C(0,\tau) = c_0, \quad \tau(0) = a. \tag{6.30}$$

In this case a solution exists in the form $C = \phi(\rho)$. We can readily deduce

$$\phi(\rho) = c_0 \left\{\int_\rho^1 e^{-b\sigma^2/2} d\sigma\right\} \left\{\int_0^1 e^{-b\sigma^2/2} d\sigma\right\}^{-1}, \tag{6.31}$$

where $b = -\phi'(1)$ is a root of

$$b = c_0 e^{-b/2} \left\{\int_0^1 e^{-b\sigma^2/2} d\sigma\right\}^{-1}, \tag{6.32}$$

and $X(t)$ (or $\tau(t)$) is given by

$$X(t) = (a^2 + 2bt)^{1/2}. \tag{6.33}$$

We observe that (6.33) actually includes the well known moving boundary $X(t) = (2bt)^{1/2}$ as a special case (see Hill (1987), page 12). There exists in the literature general solutions of the diffusion equation with a moving boundary of the form (6.33) and appropriate references may be found in Hill (1987). Here we simply consider the general moving boundaries $X(t)$ (see also Bluman and Cole (1974), page 235) which remain invariant under (6.18). Applications of the resulting similarity solutions involve complicated eigenfunction expansions which are beyond our scope.

When deducing the functional form of the solution $c(x,t)$ from (6.2) we require two independent integrals of the system of differential equations,

$$\frac{dx}{ds} = \xi(x,t), \quad \frac{dt}{ds} = \eta(x,t), \quad \frac{dc}{ds} = \zeta(x,t)c. \tag{6.34}$$

If we suppose that

$$\frac{dx}{dt} = \frac{\xi(x,t)}{\eta(x,t)}, \tag{6.35}$$

admits the integral $\omega(x,t) = $ constant then ω is the similarity variable and from

$$\frac{dc}{dt} = \frac{\zeta(x,t)}{\eta(x,t)}c, \tag{6.36}$$

with x replaced by $x = x(\omega,t)$ we can deduce (on treating ω as a constant) that the solution takes the functional form,

$$c(x,t) = \phi(\omega)\psi(\omega,t), \tag{6.37}$$

with ω in both ϕ and ψ being regarded as a function of x and t.

If we now consider the boundaries $x = X(t)$ left invariant by (6.1) then from $x_1 = X(t_1)$ we have

$$\frac{dX}{dt} = \frac{\xi(X,t)}{\eta(X,t)}, \tag{6.38}$$

so that the similarity variable ω defines the invariant boundaries from the equation

$$\omega(X(t),t) = \omega_0, \tag{6.39}$$

where ω_0 denotes an arbitrary constant. In particular for the classical group we can deduce from $(6.18)_1$, $(6.18)_2$ and (6.38)

$$\frac{d}{dt}\left\{\frac{X}{(\alpha + 2\beta t + \gamma t^2)^{1/2}}\right\} = \frac{\kappa + \delta t}{(\alpha + 2\beta t + \gamma t^2)^{3/2}}. \tag{6.40}$$

In the integration of (6.40) there are four cases which must be considered separately.

Case (i) $\beta^2 \neq \alpha\gamma$. In this case we find that the invariant boundary takes the form

$$X(t) = At + B + \omega_0(\alpha + 2\beta t + \gamma t^2)^{1/2}, \tag{6.41}$$

where the constants A and B are defined by

$$A = \left(\frac{\kappa\gamma - \delta\beta}{\alpha\gamma - \beta^2}\right), \quad B = \left(\frac{\kappa\beta - \delta\alpha}{\alpha\gamma - \beta^2}\right). \tag{6.42}$$

We observe that (6.41) contains (6.33) as a special case and that since it contains six arbitrary constants α, β, γ, δ, κ and ω_0 it may perhaps be utilized as an approximate expression for an unknown moving boundary.

Linear Partial Differential Equations

Case (ii) $\beta^2 = \alpha\gamma$, $\gamma \neq 0$. For this case we have

$$X(t) = \frac{(\delta\beta - \kappa)}{2(t+\beta)} + \omega_0(t+\beta) - \delta. \tag{6.43}$$

Case (iii) $\beta = \gamma = 0$, $\alpha \neq 0$. In this case we find

$$X(t) = \kappa t + \frac{\delta t^2}{2} + \omega_0. \tag{6.44}$$

Case (iv) $\alpha = \beta = \gamma = 0$, $\delta \neq 0$. For this case the similarity variable is simply $\omega = t$ and the invariant boundaries are therefore $t =$ constant.

In Bluman and Cole (1974)(page 235) the above moving boundaries are exploited for an *inverse* moving boundary problem in the sense that the heat input on the moving boundary is not prescribed but rather determined so as to be consistent with the assumed special form of moving boundary. The resulting solutions of (6.3) corresponding to the above four cases are outlined in Problems 6, 7, 8 and 9.

6.6 Fokker-Planck equation

In this section we consider classical groups of the form (6.1) which leave the Fokker-Planck equation (6.4) invariant. Bluman and Cole (1974) (page 258) present a detailed analysis of a boundary value problem for the special case of (6.4) with $p(x)$ equal to a constant. Here we first obtain the general form of the group for arbitrary $p(x)$ and $q(x)$. We then illustrate these results for particular functions $p(x)$ and $q(x)$. It is convenient here to introduce the functions $I(x)$ and $J(x)$ which we define by

$$I(x) = \int^x \frac{dy}{p(y)^{1/2}}, \quad J(x) = \frac{d}{dx}\left(p(x)^{1/2}\right) + \frac{q(x)}{p(x)^{1/2}}. \tag{6.45}$$

These are well defined provided $p(x)$ is non-zero in the interval under consideration. Further thoughout this section primes denote differentiation with respect to the argument indicated.

From the formulae for the transformed partial derivatives (6.8), (6.10) and (6.13) and making use of (6.4) to eliminate $\partial^2 c/\partial x^2$ we find that the condition for invariance of (6.4) becomes

$$c\frac{\partial \zeta}{\partial t} + \left(\zeta - \frac{\partial \eta}{\partial t}\right)\frac{\partial c}{\partial t} - \frac{\partial \xi}{\partial t}\frac{\partial c}{\partial x}$$

$$= \left(\zeta - 2\frac{\partial \xi}{\partial x} + \frac{p'(x)}{p(x)}\xi\right)\left\{\frac{\partial c}{\partial t} - (p'(x) + q(x))\frac{\partial c}{\partial x} - q'(x)c\right\}$$

$$+ p(x)\left\{c\frac{\partial^2 \zeta}{\partial x^2} + \left(2\frac{\partial \zeta}{\partial x} - \frac{\partial^2 \xi}{\partial x^2}\right)\frac{\partial c}{\partial x} - \frac{\partial^2 \eta}{\partial x^2}\frac{\partial c}{\partial t} - 2\frac{\partial \eta}{\partial x}\frac{\partial^2 c}{\partial x \partial t}\right\}$$

$$+ (p'(x) + q(x))\left\{c\frac{\partial \zeta}{\partial x} + \left(\zeta - \frac{\partial \xi}{\partial x}\right)\frac{\partial c}{\partial x} - \frac{\partial \eta}{\partial x}\frac{\partial c}{\partial t}\right\}$$

$$+ (p''(x) + q'(x))\xi\frac{\partial c}{\partial x} + c(\xi q''(x) + \zeta q'(x)). \tag{6.46}$$

From the coefficient of $\partial^2 c/\partial x \partial t$ we have immediately that $\eta = \eta(t)$, that is a function of t only. From the coefficient of $\partial c/\partial t$ it is a simple matter to deduce that ξ is given by

$$\xi(x, t) = \frac{\eta'(t)}{2}p(x)^{1/2}I(x) + \rho(t)p(x)^{1/2}, \tag{6.47}$$

where $\rho(t)$ denotes an arbitrary function of t and $I(x)$ is the indefinite integral defined by $(6.45)_1$. From the coefficient of $\partial c/\partial x$ in (6.46) we obtain

$$\zeta(x, t) = -\frac{\eta''(t)}{8}I^2 - \frac{\rho'(t)}{2}I - \frac{\eta'(t)}{4}IJ - \frac{\rho(t)}{2}J + \sigma(t), \tag{6.48}$$

where $\sigma(t)$ denotes a further arbitrary function of t and $J(x)$ is defined by $(6.45)_2$. On substituting the above expressions for ξ, η and ζ into the equation obtained by equating the coefficient of c in (6.46) to zero we obtain the equation,

$$\frac{\eta'''(t)}{8}I(x)^2 + \frac{\rho''(t)}{2}I(x) - \frac{\eta''(t)}{4} - \sigma'(t)$$

$$= \frac{\eta'(t)}{4}\left\{\frac{p(x)^{1/2}}{2}\phi'(x)I(x) + \phi(x)\right\} + \frac{\rho(t)}{4}p(x)^{1/2}\phi'(x), \tag{6.49}$$

where the function $\phi(x)$ is defined by

$$\phi(x) = 2p(x)^{1/2}J'(x) + J(x)^2 - 4q'(x). \tag{6.50}$$

In the analysis of (6.49) there are two distinct cases to consider.

Linear Partial Differential Equations

Case (i) $\rho(t)$ non-zero. In this case

$$\phi(x) = C_1 I(x)^2 + 2C_2 I(x) + C_3, \qquad (6.51)$$

where C_1, C_2 and C_3 denote arbitrary constants and the functions $\eta(t)$, $\rho(t)$ and $\sigma(t)$ are obtained by solving the following equations,

$$\begin{aligned} \eta'''(t) - 4C_1 \eta'(t) &= 0, \\ \rho''(t) - C_1 \rho(t) &= \frac{3C_2}{2} \eta'(t), \\ \sigma'(t) &= \frac{-\eta''(t)}{4} - C_3 \frac{\eta'(t)}{4} - C_2 \frac{\rho(t)}{2}. \end{aligned} \qquad (6.52)$$

Case (ii) $\rho(t)$ zero. In this case we have

$$\frac{p(x)^{1/2}}{2} \phi'(x) I(x) + \phi(x) = 2C_1 I(x)^2 + C_3,$$

which upon integration gives

$$\phi(x) = C_1 I(x)^2 + C_3 + \frac{C_4}{I(x)^2}, \qquad (6.53)$$

where C_4 denotes a further arbitrary constant. For this case the functions $\eta(t)$ and $\sigma(t)$ are determined by solving the following equations,

$$\begin{aligned} \eta'''(t) - 4C_1 \eta'(t) &= 0, \\ \sigma'(t) &= -\frac{\eta''(t)}{4} - C_3 \frac{\eta'(t)}{4}. \end{aligned} \qquad (6.54)$$

In both cases for given $p(x)$, we obtain the most general $q(x)$ such that (6.4) admits a classical one-parameter group of transformations leaving the equation invariant, by solving

$$2p(x)^{1/2} J'(x) + J(x)^2 - 4q'(x) = f(I), \qquad (6.55)$$

where for case (i) $f(I)$ is given by

$$f(I) = C_1 I^2 + 2C_2 I + C_3, \qquad (6.56)$$

while for case (ii) we have

$$f(I) = C_1 I^2 + C_3 + \frac{C_4}{I^2}. \qquad (6.57)$$

We solve (6.55) by taking I as the indpendent variable where $I(x)$ is defined by $(6.45)_1$. From (6.55) we have

$$2\frac{dJ}{dI} + J^2 - \frac{4}{p^{1/2}}\frac{dq}{dI} = f(I), \tag{6.58}$$

and on using

$$J = \frac{d}{dI}(\log p^{1/2}) + \frac{q}{p^{1/2}},$$

equation (6.58) simplifies to yield the Ricatti equation

$$2\frac{du}{dI} + u^2 = f(I), \tag{6.59}$$

where u is defined by

$$u = \frac{d}{dI}(\log p^{1/2}) - \frac{q}{p^{1/2}}. \tag{6.60}$$

If for the time being we assume that $u = \psi(I)$ is the solution of the Ricatti equation (6.59) then from (6.60) we see that for a given function $p(x)$ the function $q(x)$ such that (6.4) admits a classical group is given by

$$q(x) = \frac{p'(x)}{2} - p(x)^{1/2}\psi(I), \quad I(x) = \int^x \frac{dy}{p(y)^{1/2}}, \tag{6.61}$$

and we observe that in general $\psi(I)$ contains four arbitrary constants.

Clearly with $f(I)$ given by either (6.56) or (6.57) equation (6.59) has a number of simple solutions for special values of the constants C_1, C_2, C_3 and C_4. For solutions of the Ricatti equation we refer the reader to Murphy (1960)(page 15). Here we give the general solution of (6.59) in terms of confluent hypergeometric functions, that is solutions of the second order differential equation

$$zw''(z) + (c - z)w'(z) - aw(z) = 0, \tag{6.62}$$

where a and c are constants (see Murphy (1960), page 331). If c is non-integer then (6.62) has linearly indpendent solutions,

$$\begin{aligned} w_1(z) &= {}_1F_1(a, c; z), \\ w_2(z) &= z^{1-c}{}_1F_1(1 + a - c, 2 - c; z), \end{aligned} \tag{6.63}$$

Linear Partial Differential Equations

where $_1F_1(a,c;z)$ is defined by

$$_1F_1(a,c;z) = \sum_{k=0}^{\infty} \frac{(a)_k}{(c)_k} \frac{z^k}{k!}, \tag{6.64}$$

where the symbol $(a)_k$ is defined by

$$\begin{aligned}(a)_0 &= 1, \\ (a)_k &= a(a+1)(a+2)\ldots(a+k-1) \quad (k \geq 1).\end{aligned} \tag{6.65}$$

We observe that if a is a negative integer then the series (6.64) reduces to a polynomial expression.

In order to reduce (6.59) to a second order linear differential equation we make the transformation

$$u(I) = \frac{2}{v(I)} \frac{dv}{dI}, \tag{6.66}$$

and (6.59) becomes

$$\frac{d^2v}{dI^2} - \frac{f(I)}{4} v = 0. \tag{6.67}$$

We consider the two cases separately.

Case (i) $\rho(t)$ non-zero and $f(I)$ given by (6.56). In this case we let

$$z = \frac{C_1^{1/2}}{2}\left(I + \frac{C_2}{C_1}\right)^2, \quad v(I) = e^{-z/2} w(z), \tag{6.68}$$

and (6.67) becomes

$$zw''(z) + \left(\frac{1}{2} - z\right) w'(z) - aw(z) = 0, \tag{6.69}$$

where the constant a is given by

$$a = \frac{1}{4} + \frac{(C_1 C_3 - C_2^2)}{8 C_1^{3/2}}. \tag{6.70}$$

Case (ii) $\rho(t)$ zero and $f(I)$ given by (6.57). In this case we let

$$z = \frac{C_1^{1/2}}{2} I^2, \quad v(I) = z^m e^{-z/2} w(z), \tag{6.71}$$

and (6.67) gives

$$zw''(z) + \left(\frac{1}{2} + 2m - z\right)w'(z) - aw(z) = 0, \tag{6.72}$$

where the constants a and m are given by

$$a = \frac{1}{4} + \frac{C_3}{8C_1^{1/2}} + m, \quad m = \frac{1}{4}\left(1 - (1 + C_4)^{1/2}\right). \tag{6.73}$$

We note that when C_4 is zero the solution for this case coincides with that for case (i) when the constant C_2 is zero.

As a simple illustration of the above consider case (i) when the constants C_1, C_2 and C_3 are such that a as given by (6.70) assumes the value $-1/2$. For case (i) we have from (6.66), (6.67) and (6.68)

$$u(I) = (2C_1^{1/2}z)^{1/2}\left(\frac{2}{w(z)}\frac{dw}{dz} - 1\right), \tag{6.74}$$

and when $a = -1/2$ the linearly independent solutions of (6.69) obtained from (6.63) are essentially (that is, apart from an arbitrary multiplicative constant)

$$w_1^*(z) = e^z - z^{1/2}\int^z \frac{e^y}{y^{1/2}}dy,$$
$$w_2^*(z) = z^{1/2}. \tag{6.75}$$

From (6.74) and (6.75) we deduce that for case (i) with $a = -1/2$ the solution of the Ricatti equation (6.59) becomes

$$u(I) = (2C_1^{1/2})^{1/2}\left\{\frac{(1-z)\left(C_5 - \int^z \frac{e^y}{y^{1/2}}dy\right) - z^{1/2}e^z}{z^{1/2}\left(C_5 - \int^z \frac{e^y}{y^{1/2}}dy\right) + e^z}\right\}, \tag{6.76}$$

where C_5 denotes a further arbitrary constant and z as a function of I is defined by $(6.68)_1$. In the following section we consider a number of special cases of (6.4).

6.7. Examples for the Fokker-Planck equation

Example 6.4 $p(x) = 1$, $q(x)$ arbitrary. In ths case we have from (6.45)

$$I(x) = x, \quad J(x) = q(x),$$

Linear Partial Differential Equations

while from (6.50) we obtain

$$\phi(x) = q(x)^2 - 2q'(x).$$

Further (6.49) becomes

$$\frac{\eta'''(t)}{8}x^2 + \frac{\rho''(t)}{2}x - \frac{\eta''(t)}{4} - \sigma'(t)$$
$$= \frac{\eta'(t)}{2}\left\{\frac{x}{2}\phi'(x) + \phi(x)\right\} + \frac{\rho(t)}{4}\phi'(x).$$

As before there are two cases to consider.

Case (i) $\rho(t)$ non-zero. In this case $q(x)$ must satisfy the Ricatti equation

$$q(x)^2 - 2q'(x) = C_1 x^2 + 2C_2 x + C_3,$$

and $\eta(t)$, $\rho(t)$ and $\sigma(t)$ are obtained from (6.52).

Case (ii) $\rho(t)$ zero. In this case

$$\frac{x}{2}\phi'(x) + \phi(x) = 2C_1 x^2 + C_3,$$

so that

$$q(x)^2 - 2q'(x) = C_1 x^2 + C_3 + \frac{C_4}{x^2},$$

and $\eta(t)$ and $\sigma(t)$ are obtained from (6.54).

[A fuller discussion and applicaiton of this example can be found in Bluman and Cole (1974), page 258.]

Example 6.5 $p(x) = 1$, $q(x) = bx$ where b is a constant. In this case from case (i) of the previous example we have

$$C_1 = b^2, \quad C_2 = 0, \quad C_3 = -2b,$$

and therefore $\eta(t)$, $\rho(t)$ and $\sigma(t)$ are determined from the equations

$$\eta'''(t) - 4b^2 \eta'(t) = 0,$$
$$\rho''(t) - b^2 \rho(t) = 0,$$
$$\sigma'(t) = -\frac{\eta''(t)}{4} + b\frac{\eta'(t)}{2}.$$

From these equations we readily deduce

$$\eta(t) = \alpha + \beta e^{2bt} + \gamma e^{-2bt},$$

$$\rho(t) = \delta e^{bt} + \kappa e^{-bt},$$

$$\sigma(t) = \gamma b e^{-2bt} + \lambda,$$

where α, β, γ, δ, κ and λ denote six arbitrary constants. Altogether using (6.47) and (6.48) we obtain

$$\xi(x,t) = bx\left(\beta e^{2bt} - \gamma e^{-2bt}\right) + \left(\delta e^{bt} + \kappa e^{-bt}\right),$$

$$\eta(x,t) = \alpha + \beta e^{2bt} + \gamma e^{-2bt},$$

$$\zeta(x,t) = \lambda + b\gamma e^{-2bt} - b\delta x e^{bt} - b^2 x^2 \beta e^{2bt}.$$

As a simple illustration consider the case $\beta = 1$ and the remaining constants zero. The global form of the one-parameter group is obtained by solving

$$\frac{dx_1}{d\epsilon} = bx_1 e^{2bt_1}, \quad \frac{dt_1}{d\epsilon} = e^{2bt_1}, \quad \frac{dc_1}{d\epsilon} = -b^2 x_1^2 e^{2bt_1} c_1,$$

subject to the usual initial conditions (6.19). We find

$$x_1 = \frac{x}{(1 - 2\epsilon b e^{2bt})^{1/2}}, \quad t_1 = t - \frac{1}{2b}\log(1 - 2\epsilon b e^{2bt}),$$

$$c_1 = c \exp\left\{-\frac{\epsilon b^2 x^2 e^{2bt}}{(1 - 2\epsilon b e^{2bt})}\right\}.$$

Further the partial differential equation (6.2) becomes

$$bx\frac{\partial c}{\partial x} + \frac{\partial c}{\partial t} = -b^2 x^2 c,$$

which has similarity variable $\omega = xe^{-bt}$ and the functional form of the solution is given by

$$c(x,t) = e^{-bx^2/2}\phi(xe^{-bt}).$$

On substituting this functional form into (6.4) with $p(x) = 1$ and $q(x) = bx$ we obtain simply

$$\phi''(\omega) = 0,$$

so that

$$c(x,t) = e^{-bx^2/2}\left(\phi_0 + \phi_1 xe^{-bt}\right),$$

Linear Partial Differential Equations

where ϕ_0 and ϕ_1 denote arbitrary constants. The more general solution types for this example are summarized in Problem 12.

Example 6.6 Consider the equation

$$\frac{\partial c}{\partial t} = a\frac{\partial^2}{\partial x^2}(xc) + b\frac{\partial}{\partial x}(xc),$$

where a and b denote arbitrary constants. In this case we have the following results:

$$p(x) = ax, \quad q(x) = a + bx,$$

$$I(x) = 2\left(\frac{x}{a}\right)^{1/2}, \quad J(x) = b\left(\frac{x}{a}\right)^{1/2} + \frac{3}{2}\left(\frac{a}{x}\right)^{1/2},$$

$$\phi(x) = \frac{b^2}{a}x + \frac{3}{4}\frac{a}{x}.$$

From these results we find from (6.49) that $\rho(t)$ is zero and $\eta(t)$ and $\sigma(t)$ are determined from

$$\eta'''(t) - b^2\eta'(t) = 0,$$

$$\sigma'(t) = -\frac{\eta''(t)}{4}.$$

Altogether we find,

$$\xi(x,t) = bx\left(\beta e^{bt} - \gamma e^{-bt}\right),$$

$$\eta(x,t) = \alpha + \beta e^{bt} + \gamma e^{-bt},$$

$$\zeta(x,t) = \lambda - b\left(\beta e^{bt} - \gamma e^{-bt}\right) - \frac{b^2}{a}x\beta e^{bt},$$

where α, β, γ and λ denote four arbitrary constants. The various solution types for this example are summarized in Problem 13.

Example 6.7 Consider the equation

$$\frac{\partial c}{\partial t} = \frac{\partial^2}{\partial x^2}(x^2 c) + b\frac{\partial}{\partial x}(xc),$$

where b is again an arbitrary constant. In this case we have

$$p(x) = x^2, \quad q(x) = (b+2)x,$$

$$I(x) = \log x, \quad J(x) = (b+3), \quad \phi(x) = (b+1)^2,$$

and (6.49) becomes,

$$\frac{\eta'''(t)}{8}(\log x)^2 + \frac{\rho''(t)}{2}\log x - \frac{\eta''(t)}{4} - \sigma'(t) = \frac{\eta'(t)}{4}(b+1)^2.$$

Thus we obtain,

$$\eta(t) = \alpha + 2\beta t + \gamma t^2,$$

$$\rho(t) = \kappa + \delta t,$$

$$\sigma(t) = -\frac{1}{2}\left(\gamma + \beta(b+1)^2\right)t - \frac{\gamma}{4}(b+1)^2 t^2 + \lambda,$$

where α, β, γ, δ, κ and λ denote six arbitrary constants. From (6.47) and (6.48) we find

$$\xi(x,t) = (\beta + \gamma t)x \log x + (\kappa + \delta t)x,$$

$$\eta(x,t) = \alpha + 2\beta t + \gamma t^2,$$

$$\zeta(x,t) = -\frac{\gamma}{4}(\log x)^2 - \frac{1}{2}\{\delta + (b+3)(\beta + \gamma t)\}\log x$$

$$-\frac{(b+1)^2}{4}(\alpha + 2\beta t + \gamma t^2) - \frac{1}{2}(\beta + \gamma t) - \frac{1}{2}(b+3)(\kappa + \delta t) + \lambda,$$

and the various solution types are summarized in Problem 14.

6.8 Non-classical groups for the diffusion equation

In this section for the diffusion equation (6.3) we make use of equation (6.2) so that the left-hand side of (6.15) depends on c only through c and $\partial c/\partial x$. We then equate the coefficients of c and $\partial c/\partial x$ to zero and obtain two equations for the determination of the non-classical groups of (6.3). For the one-parameter group (6.1) we introduce $A(x,t)$ and $B(x,t)$ defined by

$$A(x,t) = \frac{\zeta(x,t)}{\eta(x,t)}, \quad B(x,t) = \frac{\xi(x,t)}{\eta(x,t)}, \tag{6.77}$$

so that (6.2) becomes

$$\frac{\partial c}{\partial t} = Ac - B\frac{\partial c}{\partial x}. \tag{6.78}$$

On differentiating (6.78) partially with respect to x and making use of (6.3) it is a simple matter to deduce

$$\frac{\partial^2 c}{\partial x \partial t} = \left(\frac{\partial A}{\partial x} - AB\right)c + \left(A - \frac{\partial B}{\partial x} + B^2\right)\frac{\partial c}{\partial x}. \tag{6.79}$$

Linear Partial Differential Equations

On substituting (6.78) and (6.79) into (6.15) and then equating to zero the coefficients of c and $\partial c/\partial x$ then remarkably the resulting equations simplify to give,

$$\frac{\partial A}{\partial t} = \frac{\partial^2 A}{\partial x^2} - 2A\frac{\partial B}{\partial x},$$
$$\frac{\partial B}{\partial t} = \frac{\partial^2 B}{\partial x^2} - 2B\frac{\partial B}{\partial x} - 2\frac{\partial A}{\partial x}. \tag{6.80}$$

These equations determine the *non-classical group* of (6.3). Although equations (6.80) are non-linear and clearly a good deal more complicated than the original problem, we observe that any special solution of (6.80) can be employed to reduce the diffusion equation to an ordinary differential equation.

If we assume that $B = 2\partial\Phi/\partial x$ for some function $\Phi(x,t)$ then on integrating (6.80)$_2$ with respect to x and neglecting an arbitrary function of t, we have

$$A = \frac{\partial^2 \Phi}{\partial x^2} - 2\left(\frac{\partial \Phi}{\partial x}\right)^2 - \frac{\partial \Phi}{\partial t}. \tag{6.81}$$

On substituting these expressions for A and B into (6.80)$_1$ we obtain a single equation for the determination of Φ, namely

$$\frac{\partial^2 \Phi}{\partial t^2} + \frac{\partial^4 \Phi}{\partial x^4} - 2\frac{\partial^3 \Phi}{\partial t\partial x^2} - 4\frac{\partial \Phi}{\partial x}\frac{\partial^3 \Phi}{\partial x^3} - 8\left(\frac{\partial^2 \Phi}{\partial x^2}\right)^2 + 4\frac{\partial \Phi}{\partial t}\frac{\partial^2 \Phi}{\partial x^2}$$
$$+ 8\frac{\partial^2 \Phi}{\partial x^2}\left(\frac{\partial \Phi}{\partial x}\right)^2 + 4\frac{\partial \Phi}{\partial x}\frac{\partial^2 \Phi}{\partial x\partial t} = 0. \tag{6.82}$$

This equation is clearly complicated. We simply note that the classical group (6.18) arises from the case $\Phi_{xxx} = 0$ and that if $\zeta(x,t)$ is zero then

$$A(x,t) = 0, \quad B(x,t) = -\frac{1}{h}\frac{\partial h}{\partial x}, \tag{6.83}$$

where $h(x,t)$ denotes any solution of (6.3). The resulting solution of $c(x,t)$ satisfies

$$\frac{\partial c}{\partial x} = h.$$

Example 6.8 If $\Phi(x,t) = \alpha \log x$ then from (6.82) we have either $\alpha = -1/2$ or $\alpha = -3/2$.
(i) If $\alpha = -1/2$ then

$$A(x,t) = 0, \quad B(x,t) = -\frac{1}{x},$$

so that (6.78) becomes

$$\frac{\partial c}{\partial t} - \frac{1}{x}\frac{\partial c}{\partial x} = 0,$$

which has general solution

$$c(x,t) = \phi(\omega), \quad \omega = \frac{x^2}{2} + t.$$

On substituting this functional form into (6.3) we find that

$$\phi''(\omega) = 0,$$

and therefore

$$c(x,t) = \phi_0\left(\frac{x^2}{2} + t\right) + \phi_1,$$

where ϕ_0 and ϕ_1 denote arbitrary constants.

(ii) If $\alpha = -3/2$ then

$$A(x,t) = -\frac{3}{x^2}, \quad B(x,t) = -\frac{3}{x},$$

and (6.78) becomes

$$\frac{\partial c}{\partial t} - \frac{3}{x}\frac{\partial c}{\partial x} = -\frac{3}{x^2}c.$$

This equation has general solution

$$c(x,t) = x\phi(\omega), \quad \omega = \frac{x^2}{2} + 3t,$$

and from (6.3) we deduce

$$c(x,t) = \phi_0 x\left(\frac{x^2}{2} + 3t\right) + \phi_1 x,$$

where ϕ_0 and ϕ_1 denote arbitrary constants.

PROBLEMS

1. Show for the diffusion equation that the one-parameter group arising from the constants α, κ and λ in (6.18) (that is, with $\beta = \gamma = \delta = 0$) becomes

$$x_1 = x + \kappa\epsilon, \quad t_1 = t + \alpha\epsilon, \quad c_1 = e^{\lambda\epsilon}c,$$

and that the functional form of the solution of (6.3) is

$$c(x,t) = e^{\lambda x/\kappa}\phi(\alpha x - \kappa t).$$

Hence deduce the ordinary differential equation for ϕ and relate this result to Problem 14 of Chapter 5.

2. For the diffusion equation with the constants $\alpha = \gamma = \kappa = \lambda = 0$ and β and δ non-zero in (6.18), show that the global form of the one-parameter group becomes,

$$x_1 = xe^{\beta\epsilon} + t\frac{\delta}{\beta}e^{\beta\epsilon}(e^{\beta\epsilon} - 1), \quad t_1 = e^{2\beta\epsilon}t,$$

$$c_1 = c\exp\left\{-\frac{\delta}{2\beta^2}\left[(\beta x - \delta t)(e^{\beta\epsilon} - 1) + \frac{\delta}{2}t(e^{2\beta\epsilon} - 1)\right]\right\}.$$

Show also that the functional form of the solution is

$$c(x,t) = \exp\left(\frac{\delta^2 t}{4\beta^2} - \frac{\delta x}{2\beta}\right)\phi\left(\delta t^{1/2} - \frac{\beta x}{t^{1/2}}\right),$$

and obtain the resulting ordinary differential equation for ϕ.

3. With $c(x,t) = \phi(y)\psi(t)$ where $y = (x/X(t))^2$ deduce from the diffusion equation (6.3)

$$\frac{X(t)^2}{4}\frac{\dot\psi(t)}{\psi(t)} = \frac{1}{\phi(y)}\left\{y\phi''(y) + \frac{\phi'(y)}{2}(1 + yX(t)\dot X(t))\right\}.$$

Hence conclude that if $X(t)$ takes the form

$$X(t) = (\alpha + 2\beta t)^{1/2},$$

where α and β are arbitrary constants then (6.3) admits separable solutions of the form

$$c(x,t) = X(t)^{4a/\beta}\phi(y),$$

and $\phi(y)$ satisfies
$$y\phi''(y) + \left(\frac{1}{2} + \frac{\beta y}{2}\right)\phi'(y) - a\phi(y) = 0,$$
where a denotes a further arbitrary constant. Show by a simple change of independent variable that this equation reduces to the confluent hypergeometric equation.

4. With $c(x,t) = \phi(x,t)\psi(\rho,\tau)$ where
$$\rho = \frac{x}{X(t)}, \quad \frac{d\tau}{dt} = \frac{1}{X(t)^2}, \quad X(t) = (\alpha + 2\beta t + \gamma t^2)^{1/2}, \qquad (*)$$
and
$$\phi(x,t) = \frac{1}{X(t)^{1/2}} \exp\left\{-\frac{x^2 \dot{X}(t)}{4X(t)}\right\},$$
verify that the diffusion equation (6.3) simplifies to give
$$\frac{\partial \psi}{\partial \tau} = \frac{\partial^2 \psi}{\partial \rho^2} - \frac{\delta^2 \rho^2}{4}\psi,$$
where $\delta = (\beta^2 - \alpha\gamma)^{1/2}$.

5. **Continuation** With $c(x,t) = \phi(x,t)\psi(\rho,\tau)$ where ρ, τ and X are as defined by equation $(*)$ of the previous problem and $\phi(x,t)$ is given by
$$\phi(x,t) = \frac{1}{X(t)^{\nu+1}} \exp\left\{-\frac{x^2 \dot{X}(t)}{4X(t)}\right\},$$
show that the multi-dimensional diffusion equation,
$$\frac{\partial c}{\partial t} = \frac{\partial^2 c}{\partial x^2} + \frac{(2\nu+1)}{x}\frac{\partial c}{\partial x},$$
becomes
$$\frac{\partial \psi}{\partial \tau} = \frac{\partial^2 \psi}{\partial \rho^2} + \frac{(2\nu+1)}{\rho}\frac{\partial \psi}{\partial \rho} - \frac{\delta^2 \rho^2}{4}\psi,$$
where δ is as previously defined.

6. If $\beta^2 \neq \alpha\gamma$ and $\beta^2 > \alpha\gamma$ show from (6.18) that the equation corresponding to (6.37) becomes
$$c(x,t) = \frac{\phi(\omega)}{(\alpha+2\beta t+\gamma t^2)^{1/4}}\left[\frac{\beta+\gamma t - (\beta^2-\alpha\gamma)^{1/2}}{\beta+\gamma t + (\beta^2-\alpha\gamma)^{1/2}}\right]^\mu$$
$$\text{times } \exp\left\{-\frac{t}{4}(A^2+\gamma\omega^2) + \frac{A\omega}{2}(\alpha+2\beta t+\gamma t^2)^{1/2}\right\}, \qquad (**)$$

Linear Partial Differential Equations

where A is given by $(6.42)_1$ and ω and μ are defined by

$$\omega = \frac{x - (At + B)}{(\alpha + 2\beta t + \gamma t^2)^{1/2}},$$

$$\mu = \frac{1}{2(\beta^2 - \alpha\gamma)^{1/2}} \left\{ \lambda + \frac{\beta}{2} + \frac{1}{4\gamma}[\delta^2 + A^2(\alpha\gamma - \beta^2)] \right\},$$

where B is given by $(6.42)_2$. By substituting (∗∗) into the diffusion equation (6.3) deduce the following ordinary differential equation for $\phi(\omega)$,

$$\phi''(\omega) + \beta\omega\phi'(\omega) + (D\omega^2 + E)\phi(\omega) = 0,$$

where the constants D and E are given by

$$D = \frac{\alpha\gamma}{4}, \quad E = \frac{1}{4\gamma}[A^2(\beta^2 - \alpha\gamma) - \delta^2] - \lambda.$$

If $\beta^2 < \alpha\gamma$ show that in place of the square bracket in (∗∗) we have

$$\exp\left\{ (\alpha\gamma - \beta^2)^{-1/2} \left(\lambda + \frac{\beta}{2} + \frac{1}{4\gamma}[\delta^2 + A^2(\alpha\gamma - \beta^2)] \right) \tan^{-1}\left(\frac{(\alpha\gamma - \beta^2)^{1/2}}{(\beta + \gamma t)} \right) \right\}.$$

[The solutions for $\phi(\omega)$ can be expressed in terms of confluent hypergeometric functions (see Bluman and Cole (1974), page 215).]

7. If $\beta^2 = \alpha\gamma$ and $\gamma \neq 0$ deduce from (6.18) and (6.36) the following functional form

$$c(x,t) = \frac{\phi(\omega)}{(t+\beta)^{1/2}} \exp\left\{ \frac{L^2}{12(t+\beta)^3} + \frac{M}{(t+\beta)} - \frac{L\omega}{2(t+\beta)} - \frac{\omega^2(t+\beta)}{4} \right\}, \quad (***)$$

where ω, L and M are given by

$$\omega = \left\{ x + \delta + \left(\frac{\kappa - \delta\beta}{2} \right) \frac{1}{(t+\beta)} \right\} \frac{1}{(t+\beta)},$$

$$L = \frac{1}{2}(\kappa - \delta\beta), \quad M = -\frac{1}{4}(\delta^2 + 2\beta + 4\lambda).$$

Further show from (6.3) and (∗∗∗) that $\phi(\omega)$ satisfies,

$$\phi''(\omega) - (L\omega - M)\phi(\omega) = 0.$$

[The solutions of this equation can be expressed in terms of Airy functions (see Bluman and Cole (1974), page 217).]

8. If $\beta = \gamma = 0$ and $\alpha \neq 0$ show from (6.18) that (6.37) becomes

$$c(x,t) = \phi(\omega)\exp\left\{-\frac{\delta^2 t^3}{12} - \frac{\delta\kappa t^2}{4} + \lambda t - \frac{\delta}{2}\omega t\right\}, \qquad (****)$$

where ω is given by

$$\omega = x - \frac{\delta t^2}{2} - \kappa t.$$

Deduce from $(****)$ and (6.3) that $\phi(\omega)$ satisfies

$$\phi''(\omega) + \kappa\phi'(\omega) + \left(\frac{\delta\omega}{2} - \lambda\right)\phi(\omega) = 0.$$

[This equation also has solutions expressible in terms of Airy functions (see Bluman and Cole (1974), page 218).]

9. If $\alpha = \beta = \gamma = 0$ and $\delta \neq 0$ show from (6.18) that the source solution results, namely

$$c(x,t) = \frac{\phi_0}{(t+\kappa)^{1/2}}\exp\left\{-\frac{(x-2\lambda)^2}{4(t+\kappa)}\right\},$$

where ϕ_0 denotes an arbitrary constant.

10. Show that the equation

$$\frac{\partial c}{\partial t} = \frac{\partial}{\partial x}\left\{p(x)\frac{\partial c}{\partial x}\right\},$$

can be transformed to the classical diffusion equation

$$\frac{\partial C}{\partial t} = \frac{\partial^2 C}{\partial y^2},$$

by means of the transformation,

$$c(x,t) = \alpha(x)C(y,t), \quad y = \beta(x),$$

if and only if $p(x)$ takes the form,

$$p(x) = (C_1 x + C_2)^{4/3},$$

where C_1 and C_2 denote arbitrary constants.

Linear Partial Differential Equations

11. Observe from Example 6.5 that the similarity variable ω for the classical group of

$$\frac{\partial c}{\partial t} = \frac{\partial^2 c}{\partial x^2} + b\frac{\partial}{\partial x}(xc),$$

is obtained by integrating,

$$\frac{d}{dt}\left\{\frac{x}{(\alpha + \beta e^{2bt} + \gamma e^{-2bt})^{1/2}}\right\} = \frac{(\delta e^{bt} + \kappa e^{-bt})}{(\alpha + \beta e^{2bt} + \gamma e^{-2bt})^{3/2}}, \quad (+)$$

where the constants α, β, γ, δ and κ are the same as those used in Example 6.5. By making the transformation

$$\rho = e^{2bt},$$

show that the integration of $(+)$ can be effected by considering separately four distinct cases in a completely analogous manner to the corresponding integration for the classical diffusion equation. Deduce in each case the similarity variable.

12. With reference to Example 6.5 establish the following solution types for which $\eta \to 0$ as $t \to 0$ and $c(x,t)$ is given by

$$c(x,t) = \phi(\omega)\psi(\omega,t).$$

(i) $\alpha = -\gamma = 1, \quad \beta = 0.$

$$\omega = \frac{x\tau + \delta^* + \kappa^*}{(\tau^2 - 1)^{1/2}} - \delta^*(\tau^2 - 1)^{1/2},$$

$$\psi(\omega,t) = \tau(\tau^2 - 1)^{\lambda^*/2}\exp\left\{-b\delta^*\left(\omega(\tau^2 - 1)^{1/2} + \frac{\delta^*\tau^2}{2}\right)\right\},$$

$$\phi''(\omega) + b\omega\phi'(\omega) - b\lambda^*\phi(\omega) = 0.$$

(ii) $\beta = -\alpha = 1, \quad \gamma = 0.$

$$\omega = \frac{x + (\delta^* + \kappa^*)\tau}{(\tau^2 - 1)^{1/2}} + \kappa^*\frac{(\tau^2 - 1)^{1/2}}{\tau},$$

$$\psi(\omega,t) = \left(\frac{\tau^2 - 1}{\tau^2}\right)^{\lambda^*}\exp\left\{-\frac{b}{2}\left[(\omega^2 + (\delta^* + 2\kappa^*)^2)\tau^2 - 2(\delta^* + 2\kappa^*)\omega\tau(\tau^2 - 1)^{1/2}\right]\right\},$$

$$\phi''(\omega) - b\omega\phi'(\omega) - (2\lambda^* + 1)\phi(\omega) = 0.$$

(iii) $\beta = -\gamma = 1, \quad \alpha = 0$.

$$\omega = \frac{\tau x + \delta^* + \kappa * \tau^2}{(\tau^4 - 1)^{1/2}},$$

$$\psi(\omega, t) = \frac{\tau}{(\tau^4 - 1)^{1/4}} \left| \frac{\tau^2 - 1}{\tau^2 + 1} \right|^{\lambda^*} \exp\left\{ -\frac{b}{2}\left[(\omega^2 + \kappa^{*2})\tau^2 - 2\kappa^*\omega(\tau^4 - 1)^{1/2}\right]\right\},$$

$$\phi''(\omega) - b(4\lambda^* + b\omega^2)\phi(\omega) = 0.$$

(iv) $\beta = \gamma = -\alpha/2 = 1$.

$$\omega = \frac{\tau x + 2\delta^*}{(\tau^2 - 1)} + \frac{(\delta^* + \kappa^*)}{(\tau^2 - 1)^2},$$

$$\psi(\omega, t) = \frac{\tau}{(\tau^2 - 1)^{1/2}} \exp\left\{ -\left[\frac{\lambda^*}{(\tau^2 - 1)} + b\left(\frac{\tau^2 \omega^2}{2} + \frac{(\delta^* + \kappa^*)\omega}{(\tau^2 - 1)} - \frac{(\delta^* + \kappa^*)^2}{6(\tau^2 - 1)^3}\right)\right]\right\},$$

$$\phi''(\omega) - 2b\phi'(\omega) - b\{(1 + 2\lambda^*) + 4(\delta^* + \kappa^*)b\omega - b\omega^2\}\phi(\omega) = 0.$$

(v) $\beta = 1, \gamma = \mu^2, \alpha = -(1 + \mu^2)$.

$$\omega = \frac{\tau x + \delta^* + \gamma_1(\tau^2 - \gamma_2)}{[(\tau^2 - 1)(\tau^2 - \mu^2)]^{1/2}},$$

$$\psi(\omega, t) = \left| \frac{\tau^2 - 1}{\tau^2 + \mu^2}\right|^{\lambda^*} \left| \frac{\tau^{2(1-\mu^2)}(\tau^2 - 1)^{\mu^2}}{\tau^2 - \mu^2}\right|^{\nu} \text{ times}$$

$$\exp\left\{ -\frac{b}{2}\left[(\omega^2 + \gamma_1^2)\tau^2 - 2\omega\gamma_1[(\tau^2 - 1)(\tau^2 - \mu^2)]^{1/2}\right]\right\},$$

$$\phi''(\omega) - (1 + \mu^2)b\omega\phi'(\omega) - b\{(1 + 2\lambda^*)(1 + \mu^2) - 2b\mu^2\omega^2\}\phi(\omega) = 0,$$

$$\gamma_1 = \frac{2(2\kappa^* + (1 + \mu^2)\delta^*)}{(1 - \mu^2)^2}, \quad \gamma_2 = \frac{(1 + \mu^2)}{2}, \quad \nu = \frac{1}{2(1 - \mu^2)}.$$

In each of the cases τ is defined by

$$\tau = e^{bt},$$

and the constants δ^*, κ^* and λ^* are appropriately redefined constants based on δ, κ and λ respectively and do not necessarily refer to the same constant in each case.

13. With reference to Example 6.6 establish the following solution types where $c(x, t)$ is given by

$$c(x, t) = \phi(\omega)\psi(\omega, t).$$

Linear Partial Differential Equations

(i) $\omega = \dfrac{x\tau}{(\tau-1)}, \quad \psi(\omega, t) = \tau(\tau-1)^{\lambda^*},$

$a\omega\phi''(\omega) + (2a + b\omega)\phi'(\omega) - b\lambda^* \phi(\omega) = 0.$

(ii) $\omega = \dfrac{x}{(\tau-1)}, \quad \psi(\omega, t) = \dfrac{(\tau-1)^{\lambda^*-1}}{\tau^{\lambda^*}} \exp\left\{-\dfrac{b\tau\omega}{a}\right\},$

$a\omega\phi''(\omega) + (2a - b\omega)\phi'(\omega) - b(\lambda^* + 1)\phi(\omega) = 0.$

(iii) $\omega = \dfrac{x\tau}{(\tau^2-1)}, \quad \psi(\omega, t) = \tau \dfrac{(\tau-1)^{\lambda^*-1}}{(\tau+1)^{\lambda^*+1}} \exp\left\{-\dfrac{b\tau^2 x}{a(\tau^2-1)}\right\},$

$a\omega\phi''(\omega) + 2a\phi'(\omega) - \left(2b\lambda^* + \dfrac{b^2}{a}\omega\right)\phi(\omega) = 0.$

(iv) $\omega = \dfrac{x\tau}{(\tau-1)^2}, \quad \psi(\omega, t) = \dfrac{\tau}{(\tau-1)^2} \exp\left\{-\dfrac{b\tau^2 x}{a(\tau-1)^2} + \dfrac{\lambda^* \tau}{(\tau-1)}\right\},$

$a\omega\phi''(\omega) + 2(a - b\omega)\phi'(\omega) - \left(\lambda^* - 2 + \dfrac{b\omega}{a}\right)\phi(\omega) = 0.$

(v) $\omega = \dfrac{x\tau}{(\tau-1)(\mu\tau-1)},$

$\psi(\omega, t) = \dfrac{\tau}{(\tau-1)(\mu\tau-1)} \left|\dfrac{\tau-1}{\mu\tau-1}\right|^{\lambda^*} \exp\left\{-\dfrac{b\mu\tau\omega}{a}\right\},$

$a\omega\phi''(\omega) + [2a - b(1+\mu)\omega]\phi'(\omega) + \left\{(1-\lambda^*) + \mu(1+\lambda^*) - \dfrac{b\mu}{a}\omega\right\}\phi(\omega) = 0.$

In each of the above cases τ is defined by

$$\tau = e^{bt},$$

and λ^* denotes an arbitrary constant.

14. With reference to Example 6.7 establish the following solution types where $c(x,t)$ is given by

$$c(x,t) = \phi(\omega)\psi(\omega, t).$$

(i) $\underline{\alpha = \beta = 0, \ \gamma = 1.}$

$\omega = \dfrac{\delta + \log x}{t} + \dfrac{\kappa}{2t^2},$

$\psi(\omega, t) = \dfrac{1}{t^{1/2}} \exp\left\{\dfrac{\kappa^2}{48t^3} - \dfrac{\lambda}{t} - \dfrac{(b+1)^2}{4}t - \dfrac{\omega^2 t}{4} - \dfrac{\omega}{4}\left[\dfrac{\kappa}{t} + 2(b+3)t\right]\right\},$

$\phi''(\omega) - \left\{\lambda + \dfrac{(b+3)}{4}\kappa + \dfrac{\kappa\omega}{2}\right\}\phi(\omega) = 0.$

(ii) $\alpha = \gamma = 0, \quad \beta = 1/2$.

$$\omega = \frac{2\kappa + \log x}{t^{1/2}} - 2\delta t^{1/2},$$

$$\psi(\omega, t) = t^\lambda \exp\left\{-\left[\delta^2 + (b+3)\delta + \frac{(b+1)^2}{4}\right]t - \frac{\omega t^{1/2}}{2}(2\delta + b + 3)\right\},$$

$$\phi''(\omega) - \frac{\omega}{2}\phi'(\omega) - \lambda\phi(\omega) = 0.$$

(iii) $\gamma = \beta = 0, \quad \alpha = 1$.

$$\omega = \log x - \kappa t - \frac{\delta t^2}{2},$$

$$\psi(\omega, t) = \exp\left\{-\left[\lambda + \frac{(b+3)}{2}\kappa + \frac{(b+3)^2}{4}\right]t\right.$$
$$\left. - \frac{\delta}{4}(\kappa + b + 3)t^2 - \frac{\delta^2 t^3}{12} - \frac{\delta\omega t}{2}\right\},$$

$$\phi''(\omega) + (b + 3 + \kappa)\phi'(\omega) + \left\{\lambda + \frac{(b+3)}{2}\kappa + \frac{(b+3)^2}{4} + \frac{\delta\omega}{2}\right\}\phi(\omega) = 0.$$

(iv) $\alpha = \mu^2, \quad \beta = 0, \quad \gamma = 1$.

$$\omega = \frac{\delta + \kappa t + \log x}{(t^2 + \mu^2)^{1/2}}, \quad P(t) = \exp\left\{\frac{1}{4}\tan^{-1}\left(\frac{t}{\mu}\right)\right\},$$

$$\psi(\omega, t) = \frac{P(t)^\lambda}{(t^2 + \mu^2)^{1/4}} \exp\left\{-\frac{1}{4}[(b+1)^2 + \kappa^2 - 2(b+3)\kappa + \omega^2]t\right.$$
$$\left. - \frac{\omega}{2}(b + 3 - \kappa)(t^2 + \mu^2)^{1/2}\right\},$$

$$\phi''(\omega) + \left(\frac{\mu^2 \omega^2}{4} - \lambda\right)\phi(\omega) = 0.$$

15. For the boundary value problem,

$$\frac{\partial c}{\partial t} = \frac{\partial}{\partial x}\left\{p(x)\frac{\partial c}{\partial x}\right\} + \frac{\partial}{\partial x}\{q(x)c\} \quad (t > 0, \; -\infty < x < \infty),$$

$$c(x, 0) = c_0 \delta(x - x_0),$$

$$c(x, t), \frac{\partial c}{\partial x}(x, t) \to 0 \text{ as } x \to \pm\infty,$$

where c_0 and x_0 denote arbitrary constants, show that the initial condition remains invariant under (6.1) provided the functions $\xi(x, t)$, $\eta(x, t)$ and $\zeta(x, t)$ satisfy

$$\xi(x_0, 0) = 0, \quad \eta(x_0, 0) = 0, \quad \zeta(x_0, 0) = -\frac{\partial \xi}{\partial x}(x_0, 0).$$

Linear Partial Differential Equations

Hence with $I_0 \equiv I(x_0)$ show that the functions $\eta(t)$, $\rho(t)$ and $\sigma(t)$ in (6.47) and (6.48) satisfy

$$\eta(0) = 0, \quad \rho(0) = -\frac{\eta'(0)}{2} I_0,$$

$$\sigma(0) = \frac{\eta''(0)}{8} I_0^2 + \frac{\rho'(0)}{2} I_0 - \frac{\eta'(0)}{2}.$$

16. **Continuation.** For case (i) of Section 6.6 show that one group leaving the boundary value problem of the previous question invariant is,

$$\xi(x,t) = 2p(x)^{1/2} \sinh(\beta t), \quad \eta(x,t) = 0,$$

$$\zeta(x,t) = \beta I_0 + (C_2/\beta)(1 - \cosh(\beta t)) - \beta I \cosh(\beta t) - J \sinh(\beta t),$$

where $\beta^2 = C_1$. Hence show that the solution takes the form

$$c(x,t) = \frac{\phi(t)v(I)}{p(x)^{1/2}} \exp\left\{-\frac{\beta I^2}{4} \coth(\beta t) + \frac{[\beta^2 I_0 + C_2(1 - \cosh(\beta t))]I}{2\beta \sinh(\beta t)}\right\},$$

where $\phi(t)$ denotes an arbitrary function of t. From the partial differential equation deduce that

$$c(x,t) = \frac{\phi_0 v(I) e^{-\alpha t}}{[p(x) \sinh(\beta t)]^{1/2}} \exp\left\{-\left[\left(\beta I + \frac{C_2}{\beta}\right)^2 + \left(\beta I_0 + \frac{C_2}{\beta}\right)^2\right] \frac{\coth(\beta t)}{4\beta} \right.$$

$$\left. + \left(\beta I + \frac{C_2}{\beta}\right)\left(\beta I_0 + \frac{C_2}{\beta}\right) \frac{1}{2\beta \sinh(\beta t)}\right\},$$

where $\alpha = (C_1 C_3 - C_2^2)/4C_1$ and ϕ_0 denotes an arbitrary constant. Show that as $t \to 0$,

$$c(x,t) \sim \frac{\phi_0 v(I) e^{-\alpha t}}{[p(x)\beta t]^{1/2}} \exp\left\{-\frac{(I - I_0)^2}{4t}\right\},$$

and observe that for given $I(x)$ the constant ϕ_0 is determined from the condition

$$\lim_{t \to 0} \int_{-\infty}^{\infty} c(x,t) dx = c_0.$$

17. **Continuation.** For case (ii) of Section 6.6 show that the functions $\eta(t)$ and $\sigma(t)$ are given by

$$\eta(t) = \sinh^2(\beta t),$$

$$\sigma(t) = \frac{\beta^2}{4} I_0^2 - \frac{\beta}{4} \sinh(2\beta t) - \frac{C_3}{4} \sinh^2(\beta t),$$

where again $\beta^2 = C_1$. Deduce from (6.2) that the functional form of the solution of the boundary value problem of Problem 15 becomes

$$c(x,t) = \frac{\phi(\omega)v(I)e^{-\gamma t}}{[p(x)\sinh(\beta t)]^{1/2}} \exp\left\{-\frac{\beta}{4}(I^2 + I_0^2)\coth(\beta t)\right\},$$

where $\omega = I(x)/\sinh(\beta t)$ is the similarity variable, $\gamma = C_3/4$ and $\phi(\omega)$ satisfies the differential equation

$$\phi''(\omega) - \left(\frac{\beta^2 I_0^2}{4} + \frac{C_4}{4\omega^2}\right)\phi(\omega) = 0.$$

With Ω and Φ defined by

$$\Omega = \beta I_0 \omega/2, \quad \phi(\omega) = \omega^{1/2} \Phi(\Omega),$$

show that

$$\Omega^2 \Phi''(\Omega) + \Omega \Phi'(\Omega) - (\Omega^2 + n^2)\Phi(\Omega) = 0,$$

where $n = (1 + C_4)^{1/2}/2$. Show further that the solution $\phi(\omega)$ giving the correct behaviour as $t \to 0$ and agreeing with the result of the previous question when $C_2 = C_4 = 0$ is given by

$$\phi(\omega) = \frac{\Phi_0 \omega^{1/2}}{2}[I_n(\Omega) + I_{-n}(\Omega)],$$

where Φ_0 is a constant and I_n denotes the usual modified Bessel function of the first kind. Verify that as $t \to 0$,

$$c(x,t) \sim \frac{\Phi_0 v(I)e^{-\gamma t}}{\beta[p(x)\pi I_0 t]^{1/2}} \exp\left\{-\frac{(I-I_0)^2}{4t}\right\}.$$

18. **Continuation.** Show that the solutions of the boundary value problem of Problem 15 for the following three equations:

(i) $\quad \dfrac{\partial c}{\partial t} = \dfrac{\partial^2 c}{\partial x^2} + b\dfrac{\partial}{\partial x}(xc),$

(ii) $\quad \dfrac{\partial c}{\partial t} = a\dfrac{\partial^2}{\partial x^2}(xc) + b\dfrac{\partial}{\partial x}(xc),$

(iii) $\quad \dfrac{\partial c}{\partial t} = \dfrac{\partial^2}{\partial x^2}(x^2 c) + b\dfrac{\partial}{\partial x}(xc),$

Linear Partial Differential Equations

where a and b denote arbitrary constants are respectively as follows,

(i) $$c(x,t) = c_0 \left\{ \frac{b}{2\pi(1-e^{-2bt})} \right\}^{1/2} \exp\left\{ -\frac{b(x-x_0 e^{-bt})^2}{2(1-e^{-2bt})} \right\},$$

(ii) $$c(x,t) = c_0 \frac{\gamma(t)}{2} \left(\frac{m(t)}{x} \right)^{1/2} I_1\left(2\gamma(t)(xm(t))^{1/2}\right) \exp\{-\gamma(t)(x+m(t))\},$$

where $m(t)$ and $\gamma(t)$ are defined by

$$m(t) = x_0 e^{-bt}, \quad \gamma(t) = \frac{b}{a(1-e^{-bt})},$$

(iii) $$c(x,t) = \frac{1}{4(\pi t)^{1/2}|x|} \exp\left\{ -\frac{[\log|x/x_0| + (b+1)t]^2}{4t} \right\}.$$

19. Deduce the one-parameter groups of the form (6.1) which leave the following Fokker-Planck equations invariant,

(i) $$\frac{\partial c}{\partial t} = \frac{\partial^2}{\partial x^2}\{(1-x^2)^2 c\},$$

(ii) $$\frac{\partial c}{\partial t} = \frac{\partial}{\partial x}\left\{\frac{x^{m+1}}{4}\frac{\partial c}{\partial x}\right\} + \frac{\partial}{\partial x}\left\{\frac{mx^m}{4}c\right\},$$

where m denotes an arbitrary constant.

20. Obtain the classical groups and resulting solutions of the following partial differential equations,

   (i) the wave equation,
   $$\frac{\partial^2 c}{\partial t^2} = \frac{\partial^2 c}{\partial x^2}.$$

   (ii) the telegrapher's equation,
   $$\sigma \frac{\partial^2 c}{\partial t^2} + \frac{\partial c}{\partial t} = \frac{\partial^2 c}{\partial x^2},$$

   where σ is a constant.

   (iii) the diffusion equation with convection,
   $$\frac{\partial c}{\partial t} + \delta \frac{\partial c}{\partial x} = \frac{\partial^2 c}{\partial x^2},$$

   where δ is a constant.

(iv) the Klein-Gordon equation,
$$\frac{\partial^2 c}{\partial t^2} = \frac{\partial^2 c}{\partial x^2} + \lambda c,$$
where λ is a constant.

(v) the Tricomi equation,
$$\frac{\partial^2 c}{\partial t^2} = t\frac{\partial^2 c}{\partial x^2}.$$

(vi) the Barenblatt equation,
$$\frac{\partial c}{\partial t} = \frac{\partial^2 c}{\partial x^2} + \alpha\frac{\partial^3 c}{\partial x^2 \partial t},$$
where α is a constant.

21. Obtain the classical groups and resulting solutions of the following partial differential equations,

    (i) Laplace's equation,
    $$\frac{\partial^2 c}{\partial x^2} + \frac{\partial^2 c}{\partial y^2} = 0.$$

    (ii) Helmholtz's equation,
    $$\frac{\partial^2 c}{\partial x^2} + \frac{\partial^2 c}{\partial y^2} + \lambda c = 0,$$

where λ is a constant.

22. The normal component of stress $\sigma(x, z)$ for soil is assumed to satisfy the following diffusion equation
$$\frac{\partial \sigma}{\partial z} = \alpha z \frac{\partial^2 \sigma}{\partial x^2},$$
with positive z vertically downwards and α is a constant. For the lower two-dimensional half-space with a concentrated load P at the origin, we require that
$$\sigma(x, 0) = P\delta(x), \quad \sigma(x, z) \to 0 \text{ as } x, z \to \infty,$$
where $\delta(x)$ is the usual Dirac delta function. Show that the appropriate solution is given by
$$\sigma(x, z) = \frac{P}{z(2\pi\alpha)^{1/2}} \exp\left(-\frac{x^2}{2\alpha z^2}\right).$$

Chapter Seven
Non-linear partial differential equations

7.1 Introduction

For non-linear partial differential equations we need to consider more general transformations which leave the given equation invariant. In this chapter for a single dependent variable c and for two independent variables x and t we consider one-parameter groups of the form

$$x_1 = f(x,t,c,\epsilon) = x + \epsilon\xi(x,t,c) + \mathbf{O}(\epsilon^2),$$
$$t_1 = g(x,t,c,\epsilon) = t + \epsilon\eta(x,t,c) + \mathbf{O}(\epsilon^2), \qquad (7.1)$$
$$c_1 = h(x,t,c,\epsilon) = c + \epsilon\zeta(x,t,c) + \mathbf{O}(\epsilon^2).$$

For known functions $\xi(x,t,c)$, $\eta(x,t,c)$ and $\zeta(x,t,c)$ the similarity variable and functional form of the solution are obtained by solving the first order partial differential equation,

$$\xi(x,t,c)\frac{\partial c}{\partial x} + \eta(x,t,c)\frac{\partial c}{\partial t} = \zeta(x,t,c). \qquad (7.2)$$

In the following section we give formulae for the infinitesimal versions of the first and second order partial derivatives of $c(x,t)$. Again for completeness we also give formulae for $\partial^2 c/\partial x \partial t$ and $\partial^2 c/\partial t^2$ although we make no use of these results. In the section thereafter we deduce the classical groups of the non-linear diffusion equation, namely

$$\frac{\partial c}{\partial t} = \frac{\partial}{\partial x}\left(D(c)\frac{\partial c}{\partial x}\right), \qquad (7.3)$$

where $D(c)$ denotes an arbitrary function of c. Equation (7.3) is a well known equation and in particular the power law diffusivities $D(c) = c^m$ have received a good deal of attention. The results of Section 7.3 are also given by Ovsjannikov (1967) and Bluman and Cole (1974) (page 295). In Section 7.4 we briefly consider the non-classical approach for the non-linear diffusion equation (7.3). Although we present no new results in this section the governing equations are summarized for the reader interested in pursuing the matter further. In Section 7.5 we give two results for equation (7.3) which although not directly related to group methods involve important transformations of non-linear diffusion equations. The first result shows that every equation of the form (7.3) can, by a sequence of transformations, be reduced to an equation of the form

$$D(x)\frac{\partial c}{\partial t} = \frac{\partial}{\partial x}\left(\frac{1}{c^2}\frac{\partial c}{\partial x}\right), \qquad (7.4)$$

134   *Differential Equations and Group Methods for Scientists and Engineers*

so that the power law diffusivity $D(c) = c^{-2}$ plays an important role. The second result is that the most general inhomogeneous and non-linear diffusion equation with diffusivity $D(x, c)$ which can be reduced by transformations to the classical linear diffusion equation (6.3), is given by equation (7.53). The final two sections of the chapter deal with similarity solutions of (7.59), (7.64) and (7.68).

7.2 Formulae for partial derivatives

In this section we deduce the infinitesimal versions of the partial derivatives $\partial c/\partial x$, $\partial c/\partial t$, $\partial^2 c/\partial x^2$, $\partial^2 c/\partial x \partial t$ and $\partial^2 c/\partial t^2$. We again use the convention that subscripts denote partial differentiation with x, t and c as three independent variables. Thus for example we have,

$$\frac{\partial \xi}{\partial x} = \xi_x + \xi_c \frac{\partial c}{\partial x}, \quad \frac{\partial \xi}{\partial t} = \xi_t + \xi_c \frac{\partial c}{\partial t}.$$

Now either directly from (7.1) or from the definition of a one-parameter group we have

$$x = x_1 - \epsilon \xi(x_1, t_1, c_1) + \mathbf{O}(\epsilon^2),$$

$$t = t_1 - \epsilon \eta(x_1, t_1, c_1) + \mathbf{O}(\epsilon^2),$$

and therefore up to order ϵ we obtain

$$\frac{\partial x}{\partial x_1} = 1 - \epsilon \left(\xi_x + \xi_c \frac{\partial c}{\partial x} \right) + \mathbf{O}(\epsilon^2), \quad \frac{\partial x}{\partial t_1} = -\epsilon \left(\xi_t + \xi_c \frac{\partial c}{\partial t} \right) + \mathbf{O}(\epsilon^2),$$

$$\frac{\partial t}{\partial x_1} = -\epsilon \left(\eta_x + \eta_c \frac{\partial c}{\partial x} \right) + \mathbf{O}(\epsilon^2), \quad \frac{\partial t}{\partial t_1} = 1 - \epsilon \left(\eta_t + \eta_c \frac{\partial c}{\partial t} \right) + \mathbf{O}(\epsilon^2).$$
(7.5)

First for $\partial c/\partial x$ we have

$$\frac{\partial c_1}{\partial x_1} = \frac{\partial c_1}{\partial x} \frac{\partial x}{\partial x_1} + \frac{\partial c_1}{\partial t} \frac{\partial t}{\partial x_1},$$

and using (7.1) and (7.5) we obtain

$$\frac{\partial c_1}{\partial x_1} = \left\{ \frac{\partial c}{\partial x} + \epsilon \left(\zeta_x + \zeta_c \frac{\partial c}{\partial x} \right) \right\} \left\{ 1 - \epsilon \left(\xi_x + \xi_c \frac{\partial c}{\partial x} \right) \right\}$$

$$+ \left\{ \frac{\partial c}{\partial t} + \epsilon \left(\zeta_t + \zeta_c \frac{\partial c}{\partial t} \right) \right\} \left\{ -\epsilon \left(\eta_x + \eta_c \frac{\partial c}{\partial x} \right) \right\} + \mathbf{O}(\epsilon^2),$$

which simplifies to give

$$\frac{\partial c_1}{\partial x_1} = \frac{\partial c}{\partial x} + \epsilon \left\{ \zeta_x + (\zeta_c - \xi_x) \frac{\partial c}{\partial x} - \eta_x \frac{\partial c}{\partial t} - \xi_c \left(\frac{\partial c}{\partial x} \right)^2 - \eta_c \frac{\partial c}{\partial t} \frac{\partial c}{\partial x} \right\} + \mathbf{O}(\epsilon^2). \quad (7.6)$$

Similarly for $\partial c/\partial t$ we have

$$\frac{\partial c_1}{\partial t_1} = \frac{\partial c_1}{\partial x}\frac{\partial x}{\partial t_1} + \frac{\partial c_1}{\partial t}\frac{\partial t}{\partial t_1},$$

so that from (7.1) and (7.5) we obtain

$$\frac{\partial c_1}{\partial t_1} = \left\{\frac{\partial c}{\partial x} + \epsilon\left(\zeta_x + \zeta_c\frac{\partial c}{\partial x}\right)\right\}\left\{-\epsilon\left(\xi_t + \xi_c\frac{\partial c}{\partial t}\right)\right\}$$
$$+ \left\{\frac{\partial c}{\partial t} + \epsilon\left(\zeta_t + \zeta_c\frac{\partial c}{\partial t}\right)\right\}\left\{1 - \epsilon\left(\eta_t + \eta_c\frac{\partial c}{\partial t}\right)\right\} + \mathbf{O}(\epsilon^2),$$

which becomes

$$\frac{\partial c_1}{\partial t_1} = \frac{\partial c}{\partial t} + \epsilon\left\{\zeta_t + (\zeta_c - \eta_t)\frac{\partial c}{\partial t} - \xi_t\frac{\partial c}{\partial x} - \eta_c\left(\frac{\partial c}{\partial t}\right)^2 - \xi_c\frac{\partial c}{\partial x}\frac{\partial c}{\partial t}\right\} + \mathbf{O}(\epsilon^2). \tag{7.7}$$

Again for convenience we introduce π_1 and π_2 such that

$$\pi_1 = \zeta_x + (\zeta_c - \xi_x)\frac{\partial c}{\partial x} - \eta_x\frac{\partial c}{\partial t} - \xi_c\left(\frac{\partial c}{\partial x}\right)^2 - \eta_c\frac{\partial c}{\partial t}\frac{\partial c}{\partial x},$$
$$\pi_2 = \zeta_t + (\zeta_c - \eta_t)\frac{\partial c}{\partial t} - \xi_t\frac{\partial c}{\partial x} - \eta_c\left(\frac{\partial c}{\partial t}\right)^2 - \xi_c\frac{\partial c}{\partial x}\frac{\partial c}{\partial t}, \tag{7.8}$$

so that we have simply

$$\frac{\partial c_1}{\partial x_1} = \frac{\partial c}{\partial x} + \epsilon\pi_1 + \mathbf{O}(\epsilon^2), \quad \frac{\partial c_1}{\partial t_1} = \frac{\partial c}{\partial t} + \epsilon\pi_2 + \mathbf{O}(\epsilon^2). \tag{7.9}$$

We observe that (7.8) can be written alternatively as

$$\pi_1 = \frac{\partial \zeta}{\partial x} - \frac{\partial \xi}{\partial x}\frac{\partial c}{\partial x} - \frac{\partial \eta}{\partial x}\frac{\partial c}{\partial t},$$
$$\pi_2 = \frac{\partial \zeta}{\partial t} - \frac{\partial \xi}{\partial t}\frac{\partial c}{\partial x} - \frac{\partial \eta}{\partial t}\frac{\partial c}{\partial t}. \tag{7.10}$$

For the infinitesimal version of $\partial^2 c/\partial x^2$ we have

$$\frac{\partial^2 c_1}{\partial x_1^2} = \frac{\partial}{\partial x_1}\left(\frac{\partial c_1}{\partial x_1}\right) = \frac{\partial}{\partial x}\left(\frac{\partial c_1}{\partial x_1}\right)\frac{\partial x}{\partial x_1} + \frac{\partial}{\partial t}\left(\frac{\partial c_1}{\partial x_1}\right)\frac{\partial t}{\partial x_1},$$

which using (7.5) and (7.9) gives

$$\frac{\partial^2 c_1}{\partial x_1^2} = \left\{\frac{\partial^2 c}{\partial x^2} + \epsilon\left(\pi_{1x} + \pi_{1c}\frac{\partial c}{\partial x}\right)\right\}\left\{1 - \epsilon\left(\xi_x + \xi_c\frac{\partial c}{\partial x}\right)\right\}$$
$$+ \left\{\frac{\partial^2 c}{\partial t\partial x} + \epsilon\left(\pi_{1t} + \pi_{1c}\frac{\partial c}{\partial t}\right)\right\}\left\{-\epsilon\left(\eta_x + \eta_c\frac{\partial c}{\partial x}\right)\right\} + \mathbf{O}(\epsilon^2).$$

On simplifying this result we obtain

$$\frac{\partial^2 c_1}{\partial x_1^2} = \frac{\partial^2 c}{\partial x^2} + \epsilon \left\{ \pi_{1x} + \pi_{1c}\frac{\partial c}{\partial x} - \left(\xi_x + \xi_c\frac{\partial c}{\partial x}\right)\frac{\partial^2 c}{\partial x^2} \right.$$
$$\left. - \left(\eta_x + \eta_c\frac{\partial c}{\partial x}\right)\frac{\partial^2 c}{\partial x \partial t} \right\} + \mathbf{O}(\epsilon^2). \tag{7.11}$$

From the first equation of (7.8) we have

$$\pi_{1x} = \zeta_{xx} + (\zeta_{xc} - \xi_{xx})\frac{\partial c}{\partial x} - \eta_{xx}\frac{\partial c}{\partial t} - \xi_{xc}\left(\frac{\partial c}{\partial x}\right)^2 - \eta_{xc}\frac{\partial c}{\partial t}\frac{\partial c}{\partial x}$$
$$+ (\zeta_c - \xi_x)\frac{\partial^2 c}{\partial x^2} - \eta_x\frac{\partial^2 c}{\partial x \partial t} - 2\xi_c\frac{\partial c}{\partial x}\frac{\partial^2 c}{\partial x^2} - \eta_c\frac{\partial c}{\partial t}\frac{\partial^2 c}{\partial x^2} - \eta_c\frac{\partial c}{\partial x}\frac{\partial^2 c}{\partial x \partial t}, \tag{7.12}$$
$$\pi_{1c} = \zeta_{xc} + (\zeta_{cc} - \xi_{xc})\frac{\partial c}{\partial x} - \eta_{xc}\frac{\partial c}{\partial t} - \xi_{cc}\left(\frac{\partial c}{\partial x}\right)^2 - \eta_{cc}\frac{\partial c}{\partial t}\frac{\partial c}{\partial x}.$$

Thus altogether from (7.11) and (7.12) we obtain

$$\frac{\partial^2 c_1}{\partial x_1^2} = \frac{\partial^2 c}{\partial x^2} + \epsilon \left\{ \zeta_{xx} + (2\zeta_{xc} - \xi_{xx})\frac{\partial c}{\partial x} - \eta_{xx}\frac{\partial c}{\partial t} \right.$$
$$+ (\zeta_{cc} - 2\xi_{xc})\left(\frac{\partial c}{\partial x}\right)^2 - 2\eta_{xc}\frac{\partial c}{\partial t}\frac{\partial c}{\partial x} - \xi_{cc}\left(\frac{\partial c}{\partial x}\right)^3 - \eta_{cc}\frac{\partial c}{\partial t}\left(\frac{\partial c}{\partial x}\right)^2$$
$$\left. + \left[(\zeta_c - 2\xi_x) - 3\xi_c\frac{\partial c}{\partial x} - \eta_c\frac{\partial c}{\partial t}\right]\frac{\partial^2 c}{\partial x^2} - 2\left(\eta_x + \eta_c\frac{\partial c}{\partial x}\right)\frac{\partial^2 c}{\partial x \partial t} \right\} + \mathbf{O}(\epsilon^2). \tag{7.13}$$

Similarly for $\partial^2 c/\partial x \partial t$ we have

$$\frac{\partial^2 c_1}{\partial x_1 \partial t_1} = \frac{\partial}{\partial x_1}\left(\frac{\partial c_1}{\partial t_1}\right) = \frac{\partial}{\partial x}\left(\frac{\partial c_1}{\partial t_1}\right)\frac{\partial x}{\partial x_1} + \frac{\partial}{\partial t}\left(\frac{\partial c_1}{\partial t_1}\right)\frac{\partial t}{\partial x_1},$$

and from (7.5) and (7.9)$_2$ we obtain

$$\frac{\partial^2 c_1}{\partial x_1 \partial t_1} = \frac{\partial^2 c}{\partial x \partial t} + \epsilon \left\{ \frac{\partial \pi_2}{\partial x} - \left(\xi_x + \xi_c\frac{\partial c}{\partial x}\right)\frac{\partial^2 c}{\partial x \partial t} - \left(\eta_x + \eta_c\frac{\partial c}{\partial x}\right)\frac{\partial^2 c}{\partial t^2} \right\} + \mathbf{O}(\epsilon^2). \tag{7.14}$$

Alternatively using (7.9)$_1$ we can deduce

$$\frac{\partial^2 c_1}{\partial x_1 \partial t_1} = \frac{\partial^2 c}{\partial x \partial t} + \epsilon \left\{ \frac{\partial \pi_1}{\partial t} - \left(\xi_t + \xi_c\frac{\partial c}{\partial t}\right)\frac{\partial^2 c}{\partial x^2} - \left(\eta_t + \eta_c\frac{\partial c}{\partial t}\right)\frac{\partial^2 c}{\partial x \partial t} \right\} + \mathbf{O}(\epsilon^2), \tag{7.15}$$

Non-linear Partial Differential Equations

and (7.14) and (7.15) can be shown to be the same using the expressions (7.10). From (7.8) and either of (7.14) or (7.15) we can obtain the following result,

$$\frac{\partial^2 c_1}{\partial x_1 \partial t_1} = \frac{\partial^2 c}{\partial x \partial t} + \epsilon \left\{ \zeta_{xt} + (\zeta_{ct} - \xi_{xt}) \frac{\partial c}{\partial x} + (\zeta_{cx} - \eta_{tx}) \frac{\partial c}{\partial t} \right.$$
$$- \xi_{ct} \left(\frac{\partial c}{\partial x} \right)^2 + (\zeta_{cc} - \xi_{cx} - \eta_{ct}) \frac{\partial c}{\partial x} \frac{\partial c}{\partial t} - \eta_{cx} \left(\frac{\partial c}{\partial t} \right)^2$$
$$- \xi_{cc} \left(\frac{\partial c}{\partial x} \right)^2 \frac{\partial c}{\partial t} - \eta_{cc} \left(\frac{\partial c}{\partial t} \right)^2 \frac{\partial c}{\partial x} - \left(\xi_t + \xi_c \frac{\partial c}{\partial t} \right) \frac{\partial^2 c}{\partial x^2}$$
$$\left. + \left(\zeta_c - \xi_x - \eta_t - 2\xi_c \frac{\partial c}{\partial x} - 2\eta_c \frac{\partial c}{\partial t} \right) \frac{\partial^2 c}{\partial x \partial t} - \left(\eta_x + \eta_c \frac{\partial c}{\partial x} \right) \frac{\partial^2 c}{\partial t^2} \right\} + \mathbf{O}(\epsilon^2). \quad (7.16)$$

In a similar manner we can deduce from

$$\frac{\partial^2 c_1}{\partial t_1^2} = \frac{\partial^2 c}{\partial t^2} + \epsilon \left\{ \frac{\partial \pi_2}{\partial t} - \left(\xi_t + \xi_c \frac{\partial c}{\partial t} \right) \frac{\partial^2 c}{\partial x \partial t} - \left(\eta_t + \eta_c \frac{\partial c}{\partial t} \right) \frac{\partial^2 c}{\partial t^2} \right\} + \mathbf{O}(\epsilon^2),$$

the following result

$$\frac{\partial^2 c_1}{\partial t_1^2} = \frac{\partial^2 c}{\partial t^2} + \epsilon \left\{ \zeta_{tt} + (2\zeta_{tc} - \eta_{tt}) \frac{\partial c}{\partial t} - \xi_{tt} \frac{\partial c}{\partial x} \right.$$
$$+ (\zeta_{cc} - 2\eta_{tc}) \left(\frac{\partial c}{\partial t} \right)^2 - 2\xi_{tc} \frac{\partial c}{\partial x} \frac{\partial c}{\partial t} - \eta_{cc} \left(\frac{\partial c}{\partial t} \right)^3 - \xi_{cc} \frac{\partial c}{\partial x} \left(\frac{\partial c}{\partial t} \right)^2$$
$$\left. + \left[(\zeta_c - 2\eta_t) - 3\eta_c \frac{\partial c}{\partial t} - \xi_c \frac{\partial c}{\partial x} \right] \frac{\partial^2 c}{\partial t^2} - 2 \left(\xi_t + \xi_c \frac{\partial c}{\partial t} \right) \frac{\partial^2 c}{\partial x \partial t} \right\} + \mathbf{O}(\epsilon^2). \quad (7.17)$$

We observe that (7.17) can also be deduced from (7.13) by interchanging x and t and ξ and η.

7.3 Classical groups for non-linear diffusion

Rewriting (7.3) we obtain

$$\frac{\partial c}{\partial t} = D(c) \frac{\partial^2 c}{\partial x^2} + D'(c) \left(\frac{\partial c}{\partial x} \right)^2, \quad (7.18)$$

where primes throughout denote differentiation with respect to the argument indicated. Now using

$$D(c_1) = D(c) + \epsilon \zeta D'(c) + \mathbf{O}(\epsilon^2),$$
$$D'(c_1) = D'(c) + \epsilon \zeta D''(c) + \mathbf{O}(\epsilon^2),$$

we find from (7.6), (7.7) and (7.13) that (7.18) remains invariant under (7.1) provided

$$\zeta_t + (\zeta_c - \eta_t)\frac{\partial c}{\partial t} - \xi_t\frac{\partial c}{\partial x} - \eta_c\left(\frac{\partial c}{\partial t}\right)^2 - \xi_c\frac{\partial c}{\partial x}\frac{\partial c}{\partial t}$$

$$= D(c)\left\{\zeta_{xx} + (2\zeta_{xc} - \xi_{xx})\frac{\partial c}{\partial x} - \eta_{xx}\frac{\partial c}{\partial t}\right.$$

$$+ (\zeta_{cc} - 2\xi_{xc})\left(\frac{\partial c}{\partial x}\right)^2 - 2\eta_{xc}\frac{\partial c}{\partial x}\frac{\partial c}{\partial t} - \xi_{cc}\left(\frac{\partial c}{\partial x}\right)^3$$

$$\left. - \eta_{cc}\left(\frac{\partial c}{\partial x}\right)^2\frac{\partial c}{\partial t} - 2\left(\eta_x + \eta_c\frac{\partial c}{\partial x}\right)\frac{\partial^2 c}{\partial x \partial t}\right\}$$

$$+ 2D'(c)\left\{\zeta_x + (\zeta_c - \xi_x)\frac{\partial c}{\partial x} - \eta_x\frac{\partial c}{\partial t} - \xi_c\left(\frac{\partial c}{\partial x}\right)^2 - \eta_c\frac{\partial c}{\partial x}\frac{\partial c}{\partial t}\right\}\frac{\partial c}{\partial x}$$

$$+ D''(c)\zeta\left(\frac{\partial c}{\partial x}\right)^2 + \left\{\frac{\partial c}{\partial t} - D'(c)\left(\frac{\partial c}{\partial x}\right)^2\right\}\left\{\zeta_c - 2\xi_x + \zeta\frac{D'(c)}{D(c)} - 3\xi_c\frac{\partial c}{\partial x} - \eta_c\frac{\partial c}{\partial t}\right\}, \quad (7.19)$$

where the last term, involving the two curly brackets, arises from eliminating $\partial^2 c/\partial x^2$ by means of equation (7.18). On equating the coefficients of the various partial derivatives in (7.19) to zero we obtain the following equations:

$$\frac{\partial^2 c}{\partial x \partial t}\frac{\partial c}{\partial x}, \quad \eta_c = 0,$$

$$\frac{\partial^2 c}{\partial x \partial t}, \quad \eta_x = 0,$$

$$\left(\frac{\partial c}{\partial x}\right)^3, \quad D(c)\xi_{cc} - D'(c)\xi_c = 0,$$

$$\left(\frac{\partial c}{\partial x}\right)^2\frac{\partial c}{\partial t}, \quad D(c)\eta_{cc} + D'(c)\eta_c = 0,$$

$$\left(\frac{\partial c}{\partial x}\right)^2, \quad \left[\zeta_c - 2\xi_x + \zeta\frac{D'(c)}{D(c)}\right]_c = 0,$$

$$\left(\frac{\partial c}{\partial t}\right)^2, \quad \eta_c - \eta_c = 0,$$

$$\frac{\partial c}{\partial x}\frac{\partial c}{\partial t}, \quad \xi_c + D(c)\eta_{xc} + D'(c)\eta_x = 0,$$

$$\frac{\partial c}{\partial x}, \quad \xi_t - D(c)\xi_{xx} + 2D(c)\zeta_{xc} + 2D'(c)\zeta_x = 0, \quad (7.20)$$

$$\frac{\partial c}{\partial t}, \quad \eta_t - D(c)\eta_{xx} - 2\xi_x + \zeta\frac{D'(c)}{D(c)} = 0, \quad (7.21)$$

$$c^0, \quad \zeta_t - D(c)\zeta_{xx} = 0. \quad (7.22)$$

Non-linear Partial Differential Equations

From the first seven of these equations we can readily deduce

$$\xi = \xi(x,t), \quad \eta = \eta(t), \tag{7.23}$$

$$\zeta_c + \zeta \frac{D'(c)}{D(c)} = \phi(x,t), \tag{7.24}$$

where ϕ denotes an arbitrary function of x and t. But from (7.21) and (7.24) we can deduce

$$\zeta_c = \eta'(t) - 2\frac{\partial \xi}{\partial x} + \phi(x,t),$$

and therefore $\zeta_{cc} = 0$. From (7.21) we obtain

$$\zeta = \left(\frac{D(c)}{D'(c)}\right)\left(2\frac{\partial \xi}{\partial x} - \eta'(t)\right), \tag{7.25}$$

so that either

$$2\frac{\partial \xi}{\partial x} = \eta'(t), \tag{7.26}$$

or the diffusivity $D(c)$ is such that

$$\left(\frac{D(c)}{D'(c)}\right)'' = 0,$$

that is

$$D(c) = \alpha(c+\beta)^m, \tag{7.27}$$

where α, β and m denote arbitrary constants. If (7.26) holds then from (7.20) and (7.25) we can readily deduce

$$\xi(x,t,c) = \frac{\gamma}{2}x + \kappa,$$
$$\eta(x,t,c) = \delta + \gamma t, \tag{7.28}$$
$$\zeta(x,t,c) = 0,$$

where γ, δ, κ and λ denote arbitrary constants. We note that the group (7.28) is applicable to all functions $D(c)$.

Alternatively if $D(c)$ has the form (7.27) then from (7.25) we have

$$\zeta = \frac{1}{m}(c+\beta)\left(2\frac{\partial \xi}{\partial x} - \eta'(t)\right), \tag{7.29}$$

and on substituting this expression into (7.20) we find

$$\frac{\partial \xi}{\partial t} = -D(c)\left(3 + \frac{4}{m}\right)\frac{\partial^2 \xi}{\partial x^2}, \qquad (7.30)$$

while substitution of (7.29) into (7.22) and making use of (7.30) gives

$$\eta''(t) = -8D(c)\left(1 + \frac{1}{m}\right)\frac{\partial^3 \xi}{\partial x^3}. \qquad (7.31)$$

We see that (7.30) and (7.31) give rise to two cases, namely for all constants m,

$$\frac{\partial \xi}{\partial t} = \frac{\partial^2 \xi}{\partial x^2} = \eta''(t) = 0,$$

while for $m = -4/3$ we have

$$\frac{\partial \xi}{\partial t} = \frac{\partial^3 \xi}{\partial x^3} = \eta''(t) = 0.$$

Thus for all m we have

$$\xi(x,t,c) = \kappa + \lambda x,$$
$$\eta(x,t,c) = \delta + \gamma t, \qquad (7.32)$$
$$\zeta(x,t,c) = \frac{1}{m}(c + \beta)(2\lambda - \gamma),$$

while for $m = -4/3$ we have

$$\xi(x,t,c) = \kappa + \lambda x + \mu x^2,$$
$$\eta(x,t,c) = \delta + \gamma t, \qquad (7.33)$$
$$\zeta(x,t,c) = -\frac{3}{4}(c + \beta)(4\mu x + 2\lambda - \gamma),$$

where γ, δ, κ, λ and μ denote arbitrary constants. The resulting ordinary differential equations corresponding to (7.28), (7.32) and (7.33) are given in Problems 1, 2 and 3.

Example 7.1 Deduce the source solution of the non-linear diffusion equation (7.3) with a power law diffusivity.

We need to solve

$$\frac{\partial c}{\partial t} = \frac{\partial}{\partial x}\left(c^m \frac{\partial c}{\partial x}\right), \qquad (7.34)$$

Non-linear Partial Differential Equations

such that $c(x,t)$ vanishes at infinity while initially satisfies

$$c(x,0) = c_0 \delta(x), \qquad (7.35)$$

where c_0 and m denote arbitrary constants and $\delta(x)$ is the usual Dirac delta function. Noting the elementary property of delta functions,

$$\delta(\lambda x) = \lambda^{-1} \delta(x)$$

for any non-zero constant λ, we see directly that (7.34) and (7.35) remain invariant under the one-parameter group

$$x_1 = e^\epsilon x, \quad t_1 = e^{(m+2)\epsilon} t, \quad c_1 = e^{-\epsilon} c,$$

that is,

$$\xi(x,t,c) = x, \quad \eta(x,t,c) = (m+2)t, \quad \zeta(x,t,c) = -c. \qquad (7.36)$$

This equation corresponds to (7.32) with $\beta = \delta = \kappa = 0$, $\lambda = 1$ and $\gamma = (m+2)$. We note that the more general case with $D(c)$ given by (7.27) with β non-zero appears not to admit a simple group leaving (7.35) invariant.

From (7.2) and (7.36) we see that the similarity variable and functional form of the solution are obtained by solving

$$x \frac{\partial c}{\partial x} + (m+2)t \frac{\partial c}{\partial t} = -c.$$

We find that

$$c(x,t) = \frac{\phi(\omega)}{t^n}, \quad \omega = \frac{x}{t^n}, \qquad (7.37)$$

where $n = (m+2)^{-1}$. On substituting (7.37) into (7.34) we obtain

$$\frac{d}{d\omega}\{\phi^m(\omega)\phi'(\omega) + n\omega\phi(\omega)\} = 0,$$

and since $\phi(\omega)$ vanishes at infinity, the constant of integration is zero and we obtain

$$\phi^{m-1}(\omega)\phi'(\omega) + n\omega = 0.$$

A further integration gives

$$\phi(\omega) = \left\{ C - \frac{m\omega^2}{2(m+2)} \right\}^{1/m}, \qquad (7.38)$$

where C denotes an arbitrary constant. With C given by

$$C = \frac{m\omega_1^2}{2(m+2)}, \tag{7.39}$$

we take $\phi(\omega)$ to be zero for $|\omega| > |\omega_1|$ so that the condition

$$\int_{-\infty}^{\infty} c(x,0)dx = c_0,$$

becomes

$$\int_{-\omega_1}^{\omega_1} \left(\frac{m}{2(m+2)}\right)^{1/m} (\omega_1^2 - \omega^2)^{1/m} d\omega = c_0.$$

Thus with $\omega = \omega_1 \sin\theta$ and using the formula

$$\int_{-\pi/2}^{\pi/2} (\cos\theta)^{\frac{2}{m}+1} d\theta = \frac{\sqrt{\pi}\,\Gamma(1+\frac{1}{m})}{\Gamma(\frac{3}{2}+\frac{1}{m})},$$

where $\Gamma(x)$ denotes the usual gamma function, we can readily deduce that the constant C in (7.38) is given by

$$C = \left\{\frac{m}{2(2+m)}\left(\frac{c_0}{\sqrt{\pi}}\frac{\Gamma(\frac{3}{2}+\frac{1}{m})}{\Gamma(1+\frac{1}{m})}\right)^2\right\}^{\frac{m}{(2+m)}}. \tag{7.40}$$

Thus altogether from (7.37) and (7.38) we have the the source solution of (7.34) is given by

$$c(x,t) = \frac{1}{t^n}\left\{C - \frac{mx^2}{2(m+2)t^{2n}}\right\}^{1/m}, \quad |\omega| < |\omega_1|, \tag{7.41}$$

$$c(x,t) = 0, \quad |\omega| > |\omega_1|,$$

where $n = (m+2)^{-1}$, $\omega = x/t^n$, C is given by (7.40) and ω_1 is defined by equation (7.39).

7.4 Non-classical groups for non-linear diffusion

Although there are no known non-classical groups of (7.3) we derive here the governing equations in order to illustrate the non-classical approach in a non-linear context. With $A(x,t,c)$ and $B(x,t,c)$ defined by

$$A(x,t,c) = \frac{\zeta(x,t,c)}{\eta(x,t,c)}, \quad B(x,t,c) = \frac{\xi(x,t,c)}{\eta(x,t,c)}, \tag{7.42}$$

Non-linear Partial Differential Equations

we have from (7.2)

$$\frac{\partial c}{\partial t} = A - B\frac{\partial c}{\partial x}. \tag{7.43}$$

On differentiating (7.43) partially with respect to x and using (7.18) and (7.43) to eliminate $\partial^2 c/\partial x^2$ and $\partial c/\partial t$ respectively it is a simple matter to deduce

$$\frac{\partial^2 c}{\partial t \partial x} = \left(A_x - \frac{AB}{D(c)}\right) + \left(A_c - B_x + \frac{B^2}{D(c)}\right)\frac{\partial c}{\partial x} + \left(B\frac{D'(c)}{D(c)} - B_c\right)\left(\frac{\partial c}{\partial x}\right)^2. \tag{7.44}$$

We remind the reader that the subscripts refer to partial derivatives of functions of three independent variables x, t and c. Writing

$$\theta = \frac{\partial c}{\partial x}, \tag{7.45}$$

and substituting (7.43) and (7.44) into (7.19) we obtain the following cubic expression in θ, namely

$$\zeta_t + (\zeta_c - \eta_t)(A - B\theta) - \xi_t\theta - \eta_c(A^2 - 2AB\theta + B^2\theta^2) - \xi_c\theta(A - B\theta)$$
$$= D(c)\{\zeta_{xx} + (2\zeta_{xc} - \xi_{xx})\theta - \eta_{xx}(A - B\theta) + (\zeta_{cc} - 2\xi_{xc})\theta^2$$
$$- 2\eta_{xc}\theta(A - B\theta) - \xi_{cc}\theta^3 - \eta_{cc}\theta^2(A - B\theta)\}$$
$$- 2D(c)(\eta_x + \eta_c\theta)\left\{\left(A_x - \frac{AB}{D(c)}\right) + \left(A_c - B_x + \frac{B^2}{D(c)}\right)\theta + \left(B\frac{D'(c)}{D(c)} - B_c\right)\theta^2\right\}$$
$$+ 2D'(c)\theta\{\zeta_x + (\zeta_c - \xi_x)\theta - \eta_x(A - B\theta) - \xi_c\theta^2 - \eta_c\theta(A - B\theta)\}$$
$$+ \left\{\zeta_c - 2\xi_x - 3\xi_c\theta - \eta_c(A - B\theta) + \zeta\frac{D'(c)}{D(c)}\right\}(A - B\theta - D'(c)\theta^2)$$
$$+ D''(c)\zeta\theta^2. \tag{7.46}$$

On equating to zero the coefficients of θ^3, θ^2, θ and θ^0 we obtain the following equations for the determination of $A(x,t,c)$ and $B(x,t,c)$:

$$\theta^3, \quad D(c)B_{cc} - D'(c)B_c = 0,$$
$$\theta^2, \quad \left[2B_x - A_c - A\frac{D'(c)}{D(c)}\right]_c - \frac{2BB_c}{D(c)} = 0,$$
$$\theta, \quad B_t + 2[D(c)A_x]_c - 2AB_c = D(c)B_{xx} - 2BB_x + AB\frac{D'(c)}{D(c)}, \tag{7.47}$$
$$\theta^0, \quad A_t = D(c)A_{xx} - 2AB_x + A^2\frac{D'(c)}{D(c)}.$$

We observe that with $D(c) = 1$, $A = a(x,t)c$ and $B = b(x,t)$ equations (7.47) reduce precisely to (6.80) for the two functions $a(x,t)$ and $b(x,t)$. Equations (7.47) are recorded for purposes of illustration and we make no attempt here to obtain special solutions.

7.5 Transformations of the non-linear diffusion equation

The most widely known transformation of a non-linear partial differential equation is for Burgers' equation

$$\frac{\partial u}{\partial t} + u\frac{\partial u}{\partial x} = D\frac{\partial^2 u}{\partial x^2}, \tag{7.48}$$

for which the transformation

$$u = -\frac{2D}{c}\frac{\partial c}{\partial x}, \tag{7.49}$$

reduces (7.48) to the classical diffusion equation, assuming that D is a constant. In this section we give two important results for the non-linear diffusion equation (7.3).

(i) The first result is that every non-linear diffusion equation of the form (7.3) can be transformed to the following equation with a simpler non-linearity, namely

$$D(c)\frac{\partial v}{\partial t} = v^2 \frac{\partial^2 v}{\partial c^2}, \tag{7.50}$$

where $v(c,t)$ is essentially the flux associated with equation (7.3). To see this we define $u(x,t)$ by

$$u(x,t) = D(c)\frac{\partial c}{\partial x}. \tag{7.51}$$

On multiplying (7.3) by $D(c)$ and differentiating the resulting equation partially with respect to x we find

$$\frac{\partial u}{\partial t} = \frac{\partial}{\partial x}\left(D(c)\frac{\partial u}{\partial x}\right). \tag{7.52}$$

We now introduce $v(c,t) \equiv u(x,t)$ so that (7.52) becomes

$$\frac{\partial v}{\partial c}\frac{\partial c}{\partial t} + \frac{\partial v}{\partial t} = \frac{\partial}{\partial x}\left(D(c)\frac{\partial c}{\partial x}\frac{\partial v}{\partial c}\right),$$

which on using (7.3) and (7.51) simplifies to give (7.50). We note that the equivalence of (7.4) and (7.50) is readily seen.

Non-linear Partial Differential Equations

(ii) The second result is that the most general inhomogeneous and non-linear diffusion equation with diffusivity $D(x,c)$ which can be transformed to the classical diffusion equation (6.3) takes the form

$$\frac{\partial c}{\partial t} = \frac{\partial}{\partial x}\left\{\left(\frac{\alpha x + \beta}{\gamma c + \delta}\right)^2 \frac{\partial c}{\partial x}\right\}, \tag{7.53}$$

where α, β, γ and δ denote arbitrary constants. In order to see that (7.53) can be reduced to the classical diffusion equation we can without loss of generality consider the equation,

$$\frac{\partial c}{\partial t} = \frac{\partial}{\partial x}\left\{\left(\frac{x}{c}\right)^2 \frac{\partial c}{\partial x}\right\}. \tag{7.54}$$

Instead of working with (7.54) with independent variables (x,t) we consider the same equation

$$\frac{\partial w}{\partial t} = \frac{\partial}{\partial c}\left\{\left(\frac{c}{w}\right)^2 \frac{\partial w}{\partial c}\right\}, \tag{7.55}$$

for $w(c,t)$. Making the transformation

$$w(c,t) = \frac{c}{v(c,t)}, \tag{7.56}$$

it is a simple matter to show that (7.55) becomes

$$\frac{\partial v}{\partial t} = v^2 \frac{\partial^2 v}{\partial c^2}. \tag{7.57}$$

This is clearly the same equation as (7.50) with $D(c)$ unity and therefore by introducing x such that

$$v(c,t) \equiv u(x,t) \equiv \frac{\partial c}{\partial x}, \tag{7.58}$$

equation (7.57) is equivalent to the classical diffusion equation (6.3) for $c(x,t)$.

7.6 Similarity solutions of the non-linear diffusion equation

In this section we give two examples of integrating the non-linear diffusion equation with the power law diffusivities c^{-2} and c^{-1}.

<u>Example 7.2</u> Show that the non-linear diffusion equation

$$\frac{\partial c}{\partial t} = \frac{\partial}{\partial x}\left(\frac{1}{c^2}\frac{\partial c}{\partial x}\right), \tag{7.59}$$

admits a solution of the form $c = \phi(x/t^{1/2})$ with the exact parametric respresentation

$$\omega = C_2 \left\{ e^{-u^2/4} + \frac{u}{2} \left(\int_0^u e^{-\tau^2/4} d\tau + C_1 \right) \right\},$$

$$\frac{1}{\phi} = \frac{C_2}{2} \left(\int_0^u e^{-\tau^2/4} d\tau + C_1 \right),$$

where $\omega = x/t^{1/2}$ and C_1 and C_2 denote arbitrary constants.

From $c = \phi(x/t^{1/2})$ and (7.59) we may deduce

$$\left(\frac{\phi'}{\phi^2} \right)' + \frac{\omega}{2} \phi' = 0,$$

where primes denote differentiation with respect to ω. The substitution $\psi = \phi^{-1}$ gives

$$\psi'' + \frac{\omega}{2\psi^2} \psi' = 0, \qquad (7.60)$$

and since this equation is invariant under the one-parameter group

$$\omega_1 = e^\epsilon \omega, \quad \psi_1 = e^\epsilon \psi,$$

we introduce the new variable $q = \psi/\omega$ so that $\psi = \omega q$ and we obtain,

$$\omega^2 q'' + 2\omega q' + \frac{1}{2q^2}(\omega q' + q) = 0.$$

This equation is of the Euler type and therefore we make the transformations

$$y = \log \omega, \quad p = \omega \frac{dq}{d\omega} = \frac{dq}{dy},$$

to obtain,

$$p \frac{dp}{dq} + \frac{1}{2q} + \left(1 + \frac{1}{2q^2} \right) p = 0,$$

which yet again is an Abel equation of the second kind with singular solution $p = -q$. We solve this equation in the following manner. First set $Y = p + q$ or $p = Y - q$ to give

$$\frac{dY}{dq} + \frac{1}{2q^2} + \frac{1}{2q(Y-q)} = 0,$$

and then introduce $X = q^{-1}$ so that

$$\frac{dY}{dX} = \frac{1}{2} + \frac{1}{2(XY - 1)},$$

Non-linear Partial Differential Equations

and therefore

$$\frac{dX}{dY} = 2 - \frac{2}{XY}.$$

Further the substitution $X = 2Y + u$ gives

$$\frac{du}{dY} = -\frac{2}{Y(2Y+u)},$$

or

$$\frac{dY}{du} + \frac{uY}{2} = -Y^2,$$

which at last we recognise as a standard Bernoulli which is integrated by the transformation $v = Y^{-1}$, namely

$$\frac{dv}{du} - \frac{uv}{2} = 1,$$

and therefore

$$v = e^{u^2/4}\left(\int_0^u e^{-\tau^2/4}d\tau + C_1\right).$$

Now on introducing

$$f(u) = e^{-u^2/4}\left(\int_0^u e^{-\tau^2/4}d\tau + C_1\right)^{-1},$$

we have from $v = Y^{-1} = (p+q)^{-1}$,

$$\omega\frac{dq}{d\omega} + q = f(u),$$

that is, since $\psi = \omega q$ we have

$$\frac{d\psi}{d\omega} = f(u). \tag{7.61}$$

Further from $u = X - 2Y = q^{-1} - 2(p+q)$ we obtain

$$u = \frac{\omega}{\psi} - 2\frac{d\psi}{d\omega}. \tag{7.62}$$

But differentiating this equation and using (7.60) gives

$$\frac{du}{d\omega} = \frac{1}{\psi}. \tag{7.63}$$

Now from (7.61) and (7.62) we have

$$\frac{\omega}{\psi} = u + 2f(u),$$

which on using (7.63) produces the separable equation

$$\omega \frac{du}{d\omega} = u + 2f(u),$$

for which a further integration can be effected, thus

$$\frac{1/2 \left(\int_0^u e^{-\tau^2/4} d\tau + C_1 \right) du}{\{u/2 \left(\int_0^u e^{-\tau^2/4} d\tau + C_1 \right) + e^{-u^2/4}\}} = \frac{d\omega}{\omega},$$

from which we may deduce

$$\omega = C_2 \left\{ e^{-u^2/4} + \frac{u}{2} \left(\int_0^u e^{-\tau^2/4} d\tau + C_1 \right) \right\}.$$

From this equation, $\psi = \phi^{-1}$ and (7.63) we finally have

$$\frac{1}{\phi} = \frac{d\omega}{du} = \frac{C_2}{2} \left(\int_0^u e^{-\tau^2/4} d\tau + C_1 \right),$$

which agrees with the given parametric representation.

Example 7.3 Deduce an integral of the non-linear diffusion equation

$$\frac{\partial c}{\partial t} = \frac{\partial}{\partial x} \left(\frac{1}{c} \frac{\partial c}{\partial x} \right), \tag{7.64}$$

for a solution of the form $c = \phi(x/t^{1/2})$.

From the assumed form of the solution and equation (7.64) we have

$$\left(\frac{\phi'}{\phi} \right)' + \frac{\omega}{2} \phi' = 0,$$

where again $\omega = x/t^{1/2}$ and primes denote differentiation with respect to ω. On making the substitution $\psi = \log \phi$ we obtain

$$\psi'' + \frac{\omega}{2} e^{\psi} \psi' = 0, \tag{7.65}$$

and this equation is invariant under the one-parameter group

$$\omega_1 = e^{\epsilon} \omega, \quad \psi_1 = \psi - 2\epsilon,$$

so we introduce the new variable $q = \psi + 2\log\omega$ and we have

$$\omega^2 q'' + 2 + (\omega q' - 2)\frac{e^q}{2} = 0.$$

On making the transformations

$$y = \log\omega \quad p = \omega\frac{dq}{d\omega} = \frac{dq}{dy},$$

we may deduce

$$p\frac{dp}{dq} + (p-2)\left(\frac{e^q}{2} - 1\right) = 0.$$

This equation is evidently separable and therefore can be readily integrated to yield

$$(p-2)^2 e^{(p-2)} = C e^{q - e^q/2},$$

where C is a constant. Notice however, that because of the non-explicit nature of the integral it appears difficult to proceed further.

We give below an alternative approach to deducing integrals of the non-linear diffusion equation with power law diffusivity. With the assumption $c = \phi(x/t^{1/2})$ we may deduce from (7.34)

$$(\phi^m \phi')' + \frac{\omega}{2}\phi' = 0, \tag{7.66}$$

so that with $u = \phi'$ this equation becomes

$$\frac{d}{d\phi}(\phi^m u) + \frac{\omega}{2} = 0,$$

and on differentiating this equation with respect to ϕ we obtain

$$\frac{d^2}{d\phi^2}(\phi^m u) + \frac{1}{2u} = 0.$$

Since this equation remains invariant under the one-parameter group of transformations

$$\phi_1 = e^\epsilon \phi, \quad u_1 = e^{(1-m/2)\epsilon} u,$$

we select $v = u\phi^{m/2-1}$ as a new variable which gives

$$\phi^2 \frac{d^2v}{d\phi^2} + (m+2)\phi\frac{dv}{d\phi} + \frac{m(m+2)}{4}v + \frac{1}{2v} = 0. \tag{7.67}$$

Special cases of (7.67) corresponding to the above examples can be readily integrated.

(i) $m = -2$: In this case (7.67) becomes

$$\phi^2 \frac{d^2 v}{d\phi^2} + \frac{1}{2v} = 0,$$

which on making the substitutions

$$v = \phi V, \quad \phi = \frac{1}{\Phi},$$

gives

$$\frac{d^2 V}{d\Phi^2} + \frac{1}{2V} = 0.$$

On multiplying by $dV/d\Phi$ and integrating we obtain

$$\left(\frac{dV}{d\Phi}\right)^2 + \log V = C_1,$$

so that

$$\Phi = \int^{v/\phi} \frac{dV}{\sqrt{C_1 - \log V}} + C_2,$$

where C_1 and C_2 denote arbitrary constants. Thus retracing the above transformations we may obtain

$$\frac{1}{\phi} = \int^{\phi'/\phi^3} \frac{dV}{\sqrt{C_1 - \log V}} + C_2,$$

which in principle can be integrated a further time by making the substitution $\Psi = \phi^{-2}$ since then we have

$$\int^{-\Psi'/2} \frac{dV}{\sqrt{C_1 - \log V}} = \Psi^{1/2} - C_2.$$

Hence for some function f_1 we may write

$$\Psi' = f_1(\Psi^{1/2} - C_2),$$

giving finally

$$\omega + C_3 = \int^{\phi^{-2}} \frac{d\Psi}{f_1(\Psi^{1/2} - C_2)},$$

where C_3 is the third integration constant. We remind the reader that only two of C_1, C_2 and C_3 are completely arbitrary since equation (7.66) must also be satisfied (and not just its derivative).

Non-linear Partial Differential Equations

(ii) $m = -1$: In this case (7.67) becomes

$$\phi^2 \frac{d^2v}{d\phi^2} + \phi \frac{dv}{d\phi} - \frac{v}{4} + \frac{1}{2v} = 0,$$

and therefore $\Phi = \log \phi$ gives

$$\frac{d^2v}{d\Phi^2} - \frac{v}{4} + \frac{1}{2v} = 0,$$

which can be immediately integrated

$$\left(\frac{dv}{d\Phi}\right)^2 - \frac{v^2}{4} + \log v = C_1,$$

so that

$$\log \phi = \int^{\phi'/\phi^{3/2}} \frac{dv}{\sqrt{C_1 + v^2/4 - \log v}} + C_2.$$

As in case (i) a further integration can be effected by the substitution $\Psi = \phi^{-1/2}$ then we have for some function f_2

$$\Psi' = f_2(C_2 + 2\log \Psi),$$

giving finally

$$\omega + C_3 = \int^{\phi^{-1/2}} \frac{d\Psi}{f_2(C_2 + 2\log \Psi)}.$$

In both cases (i) and (ii) we have taken the positive square root in the first integral. Equally well we could have taken the negative case and for a particular problem this aspect would need to be examined more carefully. Thus although for $m = -2$ and $m = -1$ we have fully integrated the non-linear ordinary differential equations, the resulting solutions are by no means straightforward and need to be utilized with care.

7.7 High order non-linear diffusion

A number of important problems, such as the flow of a surface tension dominated thin liquid film and the diffusion of dopant in semi-conductors, give rise to a fourth order non-linear diffusion equation

$$\frac{\partial c}{\partial t} + \frac{\partial}{\partial x}\left(c^m \frac{\partial^3 c}{\partial x^3}\right) = 0, \quad (m > 0). \tag{7.68}$$

Example 7.4 Show that (7.68) admits separable solutions of the form

$$c(x,t) = \left(\frac{x^4}{m\lambda(t+t_0)}\right)^{1/m},$$

where t_0 and λ are constants such that

$$\lambda = \frac{4}{m}\left(\frac{4}{m}-1\right)\left(\frac{4}{m}-2\right)\left(\frac{4}{m}+1\right),$$

for $m \neq 2, 4$.

If we look for a solution of (7.68) of the form

$$c(x,t) = A(x)B(t),$$

then we may readily deduce the equation

$$\frac{B'(t)}{B(t)^{m+1}} + \frac{1}{A(x)}(A(x)^m A'''(x))' = 0,$$

where primes denote differentiation with respect to the indicated argument. Thus we have

$$\frac{B'(t)}{B(t)^{m+1}} = -\frac{1}{A(x)}(A(x)^m A'''(x))' = \lambda_0, \qquad (7.69)$$

where λ_0 is a constant. The first equation gives

$$\frac{B(t)^{-m}}{-m} = \lambda_0(t+t_0),$$

where t_0 is a constant, which can be rearranged to yield

$$B(t) = \left(\frac{-1}{\lambda_0 m(t+t_0)}\right)^{1/m}.$$

For $A(x)$ we assume a solution of the form

$$A(x) = \alpha x^\beta,$$

and the second half of (7.69) simplifies to give

$$\alpha^m \beta(\beta-1)(\beta-2)[(m+1)\beta - 3]x^{m\beta - 4} = -\lambda_0.$$

Non-linear Partial Differential Equations

Thus $\beta = 4/m$ and $\alpha^m = -\lambda_0/\lambda$ where λ is as defined above. Accordingly $A(x)$ becomes

$$A(x) = \left(\frac{-\lambda_0 x^4}{\lambda}\right)^{1/m},$$

from which we may deduce the given expression for $c(x,t)$.

Example 7.5 Show that the source solution of (7.68) takes the form

$$c(x,t) = \frac{1}{t^{1/(m+4)}} \phi\left(\frac{x}{t^{1/(m+4)}}\right), \qquad (7.70)$$

where $\phi(\omega)$ is even, zero outside some interval $(-\omega_1, \omega_1)$ and satisfies

$$\begin{aligned}(m+4)\phi^{m-1}\phi''' &= \omega, \\ \phi(\omega_1) = \phi'(\omega_1) &= 0,\end{aligned} \qquad (7.71)$$

where ω_1 is determined from

$$\int_{-\omega_1}^{\omega_1} \phi(\omega) d\omega = c_0,$$

and c_0 is the constant initial source strength. Further verify that the case $m = 1$ admits the solution

$$\phi(\omega) = \frac{(\omega_1^2 - \omega^2)^2}{120},$$

where $\omega_1 = (225 c_0/2)^{1/5}$.

Equation (7.68) and the initial condition

$$c(x,0) = c_0 \delta(x),$$

remain invariant under the one-parameter group of transformations

$$x_1 = e^\epsilon x, \quad t_1 = e^{(m+4)\epsilon} t, \quad c_1 = e^{-\epsilon} c,$$

from which we may deduce the functional form (7.70). On substituting (7.68) into (7.70) we obtain

$$(m+4)(\phi^m \phi''')' = (\omega \phi)',$$

where $\omega = x/t^{1/(m+4)}$. This equation can be integrated immediately to give

$$(m+4)\phi^m\phi''' - \omega\phi = C_1,$$

where C_1 is a constant. But since $\phi(\omega)$ is even we have $\phi'(0) = \phi'''(0) = 0$ and therefore C_1 is zero and we may deduce the given equation for $\phi(\omega)$.

In the case $m = 1$, this equation becomes simply

$$\phi'''(\omega) = \frac{\omega}{5},$$

which of course may be integrated to give

$$\phi(\omega) = \frac{\omega^4}{120} + C_2\omega^2 + C_3\omega + C_4.$$

Since $\phi(\omega)$ is even the constant C_3 is zero and the appropriate solution satisfying $\phi(\omega_1) = \phi'(\omega_1) = 0$ is as given. From the condition

$$\int_{-\infty}^{\infty} c(x,t)dx = \int_{-\omega_1}^{\omega_1} \phi(\omega)d\omega = c_0$$

and the given expression for $\phi(\omega)$ we may readily deduce that $\omega_1 = (225c_0/2)^{1/5}$.

Example 7.6 Show that (7.71) remains invariant under the one-parameter group

$$\omega_1 = e^\epsilon \omega, \quad \phi_1 = e^{4\epsilon/m}\phi,$$

and accordingly may be reduced to a second order differential equation.

If we look for a group of the form

$$\omega_1 = e^\epsilon \omega, \quad \phi_1 = e^{\alpha\epsilon}\phi,$$

leaving (7.71) invariant then we require

$$(m-1)\alpha + \alpha - 3 = 1,$$

or $\alpha = 4/m$. Thus we take $u(\omega) = \phi(\omega)/\omega^{4/m}$ as the new dependent variable so that

$$\phi(\omega) = \omega^\gamma u(\omega),$$

where γ denotes $4/m$. Now from the expression

$$\phi''' = \omega^\gamma u''' + 3\gamma\omega^{\gamma-1}u'' + 3\gamma(\gamma-1)\omega^{\gamma-2}u' + \gamma(\gamma-1)(\gamma-2)\omega^{\gamma-3}u,$$

we find that (7.71) becomes

$$(m+4)u^{m-1}[\omega^3 u''' + 3\gamma\omega^2 u'' + 3\gamma(\gamma-1)\omega u' + \gamma(\gamma-1)(\gamma-2)u] = 1,$$

which is of the Euler type and therefore may be simplified by the transformation

$$y = \log \omega.$$

On using the relations

$$\frac{d}{dy} = \omega \frac{d}{d\omega}, \quad \frac{d^2}{dy^2} = \omega^2 \frac{d^2}{d\omega^2} + \omega \frac{d}{d\omega},$$

$$\frac{d^3}{dy^3} = \omega^3 \frac{d^3}{d\omega^3} + 3\omega^2 \frac{d^2}{d\omega^2} + \omega \frac{d}{d\omega},$$

we may deduce

$$(m+4)u^{m-1}\left\{\frac{d^3 u}{dy^3} + 3(\gamma-1)\frac{d^2 u}{dy^2} + (3\gamma^2 - 6\gamma + 2)\frac{du}{dy} + \gamma(\gamma-1)(\gamma-2)u\right\} = 1,$$

the order of which can be reduced by one by means of the substitution $p = du/dy$ and using

$$\frac{d^2 u}{dy^2} = p\frac{dp}{du}, \quad \frac{d^3 u}{dy^3} = p\frac{d}{du}\left(p\frac{dp}{du}\right).$$

We may finally obtain the following,

$$p\frac{d^2 p}{du^2} + \left(\frac{dp}{du}\right)^2 + 3(\gamma-1)\frac{dp}{du} + (3\gamma^2 - 6\gamma + 2) + \gamma(\gamma-1)(\gamma-2)\frac{u}{p} = \frac{1}{(m+4)pu^{m-1}},$$

which although of second order appears to be a more complicated equation than the original third order equation (7.71) and further, even special values of $\gamma = 4/m$ appear not to generate solvable cases.

PROBLEMS

1. For the non-linear diffusion equation (7.3) show that the similarity variable and functional form of the solution corresponding to the group

$$\xi(x,t,c) = x + \kappa, \quad \eta(x,t,c) = 2(t+\delta), \quad \zeta(x,t,c) = 0,$$

are respectively

$$\omega = \frac{x+\kappa}{(t+\delta)^{1/2}}, \quad c = \phi(\omega).$$

Hence show that the resulting ordinary differential equation is

$$D(\phi)\phi''(\omega) + D'(\phi)\phi'(\omega)^2 + \frac{\omega}{2}\phi'(\omega) = 0.$$

2. For the non-linear diffusion equation with

$$D(c) = \alpha(c+\beta)^m,$$

show that the similarity variable and functional form of the solution corresponding to the group

$$\xi(x,t,c) = (1+\lambda)x + \kappa, \quad \eta(x,t,c) = 2(t+\delta),$$

$$\zeta(x,t,c) = \frac{2\lambda}{m}(c+\beta),$$

are given respectively by

$$\omega = \frac{(x+\frac{\kappa}{(1+\lambda)})}{(t+\delta)^{\frac{1+\lambda}{2}}}, \quad c = (t+\delta)^{\lambda/m}\phi(\omega) - \beta.$$

Show that the resulting ordinary differential equation is

$$\phi(\omega)^m \phi''(\omega) + m\phi(\omega)^{m-1}\phi'(\omega)^2 + \frac{(1+\lambda)}{2\alpha}\omega\phi'(\omega) - \frac{\lambda}{\alpha m}\phi(\omega) = 0.$$

Show that this equation can be reduced to a first order ordinary differential equation by observing that the above equation remains invariant under the one-parameter group of transformations

$$\omega_1 = e^\epsilon \omega, \quad \phi_1 = e^{2\epsilon/m}\phi.$$

3. For the non-linear diffusion equation with

$$D(c) = \alpha(c+\beta)^{-4/3},$$

consider the special case of the group

$$\xi(x,t,c) = \mu x^2 + (1+\lambda)x + \kappa, \quad \eta(x,t,c) = 2(t+\delta),$$

$$\zeta(x,t,c) = -\frac{3}{2}(c+\beta)(2\mu x + \lambda),$$

for which the constants κ, λ and μ satisfy

$$(\lambda+1)^2 = 4\mu\kappa.$$

In this case show that the similarity variable and functional form of solution are

$$\omega = (t+\delta)^{-1/2} \exp\left\{\frac{-2}{(2\mu x + \lambda + 1)}\right\},$$

$$c = -\beta + \phi(\omega)\left(x + \frac{(\lambda+1)}{2\mu}\right)^{-3} \exp\left\{\frac{-3}{(2\mu x + \lambda + 1)}\right\}.$$

Show that the resulting ordinary differential equation is

$$\phi''(\omega) - \frac{4\phi'(\omega)^2}{3\phi(\omega)} - \frac{3\phi(\omega)}{4\omega^2} + \frac{\mu^2}{2\alpha}\omega\phi(\omega)^{4/3}\phi'(\omega) = 0,$$

which can be reduced to a first order ordinary differential equation by observing that the above equation remains invariant under the group

$$\omega_1 = e^\epsilon \omega, \quad \phi_1 = e^{-3\epsilon/2}\phi.$$

4. Show that the non-classical approach applied to

$$F(x)\frac{\partial c}{\partial t} = \frac{\partial}{\partial x}\left\{\frac{1}{c^2}\frac{\partial c}{\partial x}\right\},$$

gives rise to the following four equations for the one-parameter group (7.1),

$$B_{cc} + 2\frac{B_c}{c} = 0,$$

$$\left[2B_x - A_c + \frac{2A}{c}\right]_c + c^2[F(x) - 3]BB_c = 0,$$

$$F(x)[B_t + (AB)_c] + F'(x)B^2 + \left\{2\left(\frac{A_x}{c^2}\right)_c - 3AB_c - BA_c\right\} = \frac{B_{xx}}{c^2} - 2BB_x - 2\frac{AB}{c},$$

$$F(x)[A_t + AA_c] + F'(x)AB - AA_c = \frac{A_{xx}}{c^2} - 2AB_x - 2\frac{A^2}{c},$$

where $A(x,t,c)$ and $B(x,t,c)$ are defined by (7.42) and subscripts denote partial derivatives with respect to the three independent variables x, t and c.

5. The non-linear axially symmetric diffusion or heat conduction equation in cylindrical and spherical regions can be transformed into an equation of the form

$$\frac{\partial^2 c}{\partial x^2} + \frac{k}{x}\frac{\partial c}{\partial x} = f(c)\frac{\partial c}{\partial t}, \qquad (*)$$

where $k = 1$ and $k = 2$ corresponds to cylindrical and spherical regions respectively. By considering the classical invariance of $(*)$ under the one-parameter group

$$x_1 = x + \epsilon\xi(x) + \mathbf{O}(\epsilon^2),$$
$$t_1 = t + \epsilon\eta(t) + \mathbf{O}(\epsilon^2),$$
$$c_1 = c + \epsilon\zeta(x,c) + \mathbf{O}(\epsilon^2),$$

show that for both values of k a group exists if either $f(c)$ is constant or if

$$f(c) = \alpha(c+\beta)^m,$$

where α, β and m denote arbitrary constants. In the latter situation show that for both values of k,

$$\xi(x) = x, \quad \eta(t) = (m\gamma + 2)t + \delta, \quad \zeta(x,c) = \gamma c + \lambda, \qquad (**)$$

where γ, δ and λ denote further arbitrary constants.

For the special case of $k = 1$ show that a more general group than $(**)$ exists provided

$$f(c) = \alpha(c+\beta)^{-1},$$

and deduce for example that $(*)$ admits groups of the form,

$$\xi(x) = \frac{x}{2}[\log x + (1+\gamma)], \quad \eta(t) = 2t + \delta,$$
$$\zeta(x,c) = (\log x + \gamma)(c+\beta),$$

where again γ and δ denote further arbitrary constants.

6. **Continuation.** Deduce the similarity variables, functional forms of the solutions and the resulting ordinary differential equations for the groups given in the previous problem.

7. Observe that

$$\frac{\partial^2 c}{\partial x^2} + \frac{k}{x}\frac{\partial c}{\partial x} = c^m \frac{\partial c}{\partial t}, \qquad (***)$$

remains invariant under the group,

$$x_1 = e^\epsilon x, \quad t_1 = e^{(2+mn)\epsilon} t, \quad c_1 = e^{n\epsilon} c.$$

Use this group to deduce the source solutions of $(***)$ for $k = 1$ and $k = 2$ given that for these values of k the group with $n = -2$ and $n = -3$ respectively leaves the appropriate initial condition invariant as well.

[This is because with rectangular cartesian coordinates (X, Y, Z) the appropriate initial condition for $k = 1$ is

$$c(X, Y, 0) = c_0 \delta(X) \delta(Y),$$

while for $k = 2$ we have

$$c(X, Y, Z, 0) = c_0 \delta(X) \delta(Y) \delta(Z),$$

where as usual c_0 denotes a constant specifying the strength of the source.]

8. Show that

$$\frac{\partial^2 c}{\partial x^2} + \frac{\partial^2 c}{\partial y^2} = \alpha e^{\beta c} \frac{\partial c}{\partial t}, \qquad (+)$$

remains invariant under

$$x_1 = x + \epsilon \xi_1(x, y) + \mathbf{O}(\epsilon^2),$$
$$y_1 = y + \epsilon \xi_2(x, y) + \mathbf{O}(\epsilon^2),$$
$$t_1 = t + \epsilon(\gamma t + \delta) + \mathbf{O}(\epsilon^2),$$
$$c_1 = c + \frac{\epsilon}{\beta}\left(\gamma - \frac{\partial \xi_1}{\partial x} - \frac{\partial \xi_2}{\partial y}\right) + \mathbf{O}(\epsilon^2),$$

provided $\xi_1(x, y)$ and $\xi_2(x, y)$ satisfy the Cauchy-Riemann equations, namely

$$\frac{\partial \xi_1}{\partial x} = \frac{\partial \xi_2}{\partial y}, \quad \frac{\partial \xi_1}{\partial y} = -\frac{\partial \xi_2}{\partial x},$$

and α, β, γ and δ denote arbitrary constants.

9. **Continuation.** For $\delta = 0$ and $\xi_1(x,y)$ and $\xi_2(x,y)$ given by

$$\xi_1(x,y) = x^2 - y^2, \quad \xi_2(x,y) = 2xy,$$

deduce the following similarity variables and functional form of the solution,

$$\omega = \frac{(x^2+y^2)}{y}, \quad \tau = t\exp\left\{\frac{\gamma x}{(x^2+y^2)}\right\},$$

$$c(x,y,t) = \phi(\omega,\tau) - \frac{2}{\beta}\log y - \frac{\gamma x}{\beta(x^2+y^2)}.$$

Hence from (+) deduce the following partial differential equation for $\phi(\omega,\tau)$,

$$\frac{\partial}{\partial \omega}\left(\omega^2 \frac{\partial \phi}{\partial \omega}\right) + \frac{\gamma^2 \tau}{\omega^2}\frac{\partial}{\partial \tau}\left(\tau\frac{\partial \phi}{\partial \tau}\right) = \alpha \frac{\partial \phi}{\partial \tau}e^{\beta\phi} - \frac{2}{\beta}.$$

10. **Continuation.** For $\delta = 0$ and $\xi_1(x,y)$ and $\xi_2(x,y)$ given by

$$\xi_1(x,y) = e^{nx}\cos ny, \quad \xi_2(x,y) = e^{nx}\sin ny,$$

deduce the following similarity variable and functional form of the solution,

$$\omega = e^{-nx}\sin ny, \quad \tau = t\exp\left(\frac{\gamma}{n}e^{-nx}\cos ny\right),$$

$$c(x,y,t) = \phi(\omega,\tau) - \frac{\gamma}{n\beta}e^{-nx}\cos ny - \frac{2}{\beta}\log(\sin ny).$$

Hence from (+) deduce the following partial differential equation for $\phi(\omega,\tau)$,

$$\frac{\partial^2 \phi}{\partial \omega^2} + \frac{\gamma^2 \tau}{n^2}\frac{\partial}{\partial \tau}\left(\tau\frac{\partial \phi}{\partial \tau}\right) = \frac{\alpha}{(n\omega)^2}\frac{\partial \phi}{\partial \tau}e^{\beta\phi} - \frac{2}{\beta\omega^2}.$$

11. Show that the source solution for Burgers' equation, namely

$$\frac{\partial u}{\partial t} + u\frac{\partial u}{\partial x} = D\frac{\partial^2 u}{\partial x^2},$$

$$u(x,0) = u_0\delta(x),$$

$$u(x,t) \to 0 \text{ as } x \to \pm\infty,$$

where u_0 is a constant, remains invariant under the one-parameter group

$$x_1 = e^{\epsilon}x, \quad t_1 = e^{2\epsilon}t, \quad u_1 = e^{-\epsilon}u.$$

Hence deduce that

$$u(x,t) = t^{-1/2}\phi(xt^{-1/2}),$$

and subsequently that

$$\phi(\omega) = \frac{-2De^{-\omega^2/4D}}{(C + \int_{-\infty}^{\omega} e^{-\lambda^2/4D}\,d\lambda)},$$

where $\omega = xt^{-1/2}$ and C is a constant. Deduce from the initial condition that the constant C is given by

$$C = -\frac{(\pi D)^{1/2}e^{u_0/4D}}{\sinh(u_0/4D)}.$$

12. Show that the classical groups of the non-linear wave equation,

$$\frac{\partial^2 c}{\partial t^2} = f(c)^2 \frac{\partial^2 c}{\partial x^2} \qquad (++)$$

where $f(c)$ is non-constant are summarized by the following three cases:

(i) $\underline{f(c) \text{ arbitrary}}$

$$\xi(x,t,c) = \gamma x + \delta,$$
$$\eta(x,t,c) = \gamma t + \kappa,$$
$$\zeta(x,t,c) = 0.$$

(ii) $\underline{f(c) = \alpha(c+\beta)^m}$

$$\xi(x,t,c) = \gamma x + \delta + \lambda m x,$$
$$\eta(x,t,c) = \gamma t + \kappa,$$
$$\zeta(x,t,c) = \lambda(c+\beta).$$

(iii) $\underline{f(c) = \alpha(c+\beta)^2}$

$$\xi(x,t,c) = \gamma x + \delta + 2\lambda x + \mu x^2,$$
$$\eta(x,t,c) = \gamma t + \kappa,$$
$$\zeta(x,t,c) = (\lambda + \mu x)(c+\beta).$$

In each case deduce the similarity variables, functional forms of the solution and the resulting ordinary differential equations.

13. **Continuation.** The fundamental solution of the non-linear wave equation (++) of the previous problem satisfies the initial data

$$c(x,0) = 0, \quad \frac{\partial c}{\partial t}(x,0) = \delta(x). \qquad (+++)$$

Show that for all wave speeds $f(c)$ the fundamental solution remains invariant under the one-parameter group

$$x_1 = e^\epsilon x, \quad t_1 = e^\epsilon t, \quad c_1 = c,$$

and hence takes the form

$$c(x,t) = \phi(xt^{-1}).$$

Deduce from (++) that ϕ satisfies the ordinary differential equation

$$\omega^2 \phi''(\omega) + 2\omega \phi'(\omega) = f(\phi)^2 \phi''(\omega),$$

where $\omega = xt^{-1}$.

14. **Continuation.** For the linear case $f(c) = f_0$ where f_0 is a constant deduce that $\phi'(\omega)$ is given by

$$\phi'(\omega) = \frac{C_1}{(f_0^2 - \omega^2)},$$

where C_1 denotes an arbitrary constant. Hence show that

$$c(x,t) = \frac{C_1}{2f_0} \log \left| \frac{x + f_0 t}{x - f_0 t} \right| + C_2,$$

where C_2 denotes a further arbitrary constant. This solution becomes infinite at $x = f_0 t$ unless $C_1 = 0$, in which case $c(x,t) = C_2$ and the appropriate solution is $c(x,t) = (2f_0)^{-1}$ for $|x| < f_0 t$ and zero otherwise.

15. **Continuation.** For the case $f(c) = c$ show that the ordinary differential equation of Problem 13 remains invariant under the one-parameter group

$$\omega_1 = e^\epsilon \omega, \quad \phi_1 = e^\epsilon \phi.$$

Hence with,

$$\psi = \frac{\phi}{\omega}, \quad \tau = \log \omega, \quad p = \frac{d\psi}{d\tau},$$

Non-linear Partial Differential Equations

deduce the Abel equation of the second kind,

$$p\frac{dp}{d\psi} + p\frac{(3-\psi^2)}{(1-\psi^2)} + \frac{2\psi}{(1-\psi^2)} = 0.$$

16. Obtain the classical groups and resulting solutions of the following non-linear partial differential equations,

   (i) the non-linear Burger's equation,

   $$\frac{\partial u}{\partial t} + u\frac{\partial u}{\partial x} = \frac{\partial^2 u}{\partial x^2}.$$

   (ii) the Korteweg-de Vries equation,

   $$\frac{\partial u}{\partial t} + u\frac{\partial u}{\partial x} = \frac{\partial^3 u}{\partial x^3}.$$

17. Show that the non-linear diffusion equation with power law diffusivity,

$$\frac{\partial c}{\partial t} = \frac{\partial}{\partial x}\left(c^m \frac{\partial c}{\partial x}\right),$$

remains invariant under the one-parameter group

$$x_1 = e^\epsilon x, \quad t_1 = e^{\alpha\epsilon}t, \quad c_1 = e^{\beta\epsilon}c,$$

provided that $\alpha + m\beta = 2$. Hence deduce that this equation admits solutions of the form

$$c(x,t) = t^{\beta/\alpha}\phi\left(\frac{x}{t^{1/\alpha}}\right).$$

18. **Continuation.** The similarity source solution of this non-linear diffusion equation as given in Example 7.1 arises from the case $\alpha = m+2$ and $\beta = -1$ and takes the form

$$c(x,t) = \frac{1}{t^{1/(m+2)}}\left(C - \frac{mx^2}{2(m+2)t^{2/(m+2)}}\right)^{1/m},$$

where C is a constant. Show that in the limit as C tends to zero, this solution becomes

$$c(x,t) = \left(-\frac{mx^2}{2(m+2)t}\right)^{1/m},$$

which is another similarity solution, arising from the case $\alpha = 2$ and $\beta = 0$. On translating t, observe that this latter solution corresponds to the so-called 'waiting time solution'

$$c(x,t) = \left(\frac{mx^2}{2(m+2)(t_0-t)}\right)^{1/m},$$

where t_0 denotes an arbitrary constant. Verify that this waiting time solution can also be deduced by the technique of separation of variables.

19. Show that

$$2\phi\phi'' - \phi'^2 + \frac{1}{6}(\omega_1^2 - \omega^2) = 0,$$

is the appropriate integral of (7.71) for the special case of $m = 2$.

20. Show that

$$\frac{\partial c}{\partial t} = (-1)^n \frac{\partial}{\partial x}\left(c^m \frac{\partial^{2n+1} c}{\partial x^{2n+1}}\right),$$

admits separable solutions of the form

$$c(x,t) = \left(\frac{x^{2(n+1)}}{\lambda m[t_0 + (-1)^{n+1} t]}\right)^{1/m},$$

where t_0 and λ are constants such that

$$\lambda = \frac{2(n+1)}{m}\left(\frac{2(n+1)}{m} - 1\right)\left(\frac{2(n+1)}{m} - 2\right)\ldots\left(\frac{2(n+1)}{m} - 2n\right)\left(\frac{2(n+1)}{m} + 1\right),$$

for those values of m and n for which λ is non-zero.

21. **Continuation.** Show that the similarity source solution for the equation of the previous problem takes the form

$$c(x,t) = \frac{1}{t^k}\phi\left(\frac{x}{t^k}\right),$$

where $k = (m + 2n + 2)^{-1}$. With $\omega = x/t^k$ deduce that

$$\phi^{m-1}\frac{d^{2n+1}\phi}{d\omega^{2n+1}} = (-1)^{n+1} k\omega,$$

where $\phi(\omega)$ is even, zero outside $(-\omega_1, \omega_1)$ and such that

$$\phi(\omega_1) = \phi'(\omega_1) = \phi''(\omega_1) = \ldots = \phi^{(n)}(\omega_1) = 0,$$

Non-linear Partial Differential Equations

where ω_1 is determined from the condition

$$\int_{-\omega_1}^{\omega_1} \phi(\omega)\,d\omega = c_0,$$

and as usual c_0 is the constant initial source strength.

22. **Continuation.** For the special case of $m = 1$, verify that the appropriate similarity source solution is given by

$$c(x,t) = \frac{1}{t^{1/(2n+3)}} \phi\left(\frac{x}{t^{1/(2n+3)}}\right),$$

where $\phi(\omega)$ is given by

$$\phi(\omega) = \frac{(\omega_1^2 - \omega^2)^{n+1}}{(2n+3)!},$$

and ω_1 is determined from

$$\omega_1 = \left(\frac{(2n+3)!\,\Gamma(n+5/2)c_0}{\sqrt{\pi}(n+1)!}\right)^{1/(2n+3)}.$$

23. For the Korteweg-de Vries equation given in Problem 16 show that,

   (i) the equation remains invariant under the one-parameter group

$$x_1 = e^\epsilon x, \quad t_1 = e^{3\epsilon}t, \quad u_1 = e^{-2\epsilon}u,$$

and the resulting similarity solution,

$$u(x,t) = t^{-2/3}\phi(x/t^{1/3}),$$

satisfies the differential equation

$$\phi''' = \phi\phi' - (\omega\phi' + 2\phi)/3,$$

where $\omega = x/t^{1/3}$.

   (ii) the equation remains invariant under the one-parameter group

$$x_1 = x + \epsilon t, \quad t_1 = t, \quad u_1 = u + \epsilon,$$

and the resulting solution is simply

$$u(x,t) = (x - x_0)/t,$$

where x_0 is a constant.

BIBLIOGRAPHY

Bluman, G.W. and Cole, J.D., Similarity methods for differential equations, Applied Mathematical Sciences 13, Springer-Verlag, New York, 1974.

Chester, W., Continuous transformations and differential equations, Journal of the Institute of Mathematics and its Applications 19, 343-386, 1977.

Cohen, A., An introduction to the Lie theory of one-parameter groups with applications to the solution of differential equations, D.C. Heath and Co. Publishers, New York, 1911.

Coppel, W.A., Disconjugacy, Lecture Notes in Mathematics 220, Springer-Verlag, New York, 1971.

Dickson, L.E., Differential equations from the group standpoint, Annals of Mathematics 25, 287-378, 1924.

Gröbner, W. and Knapp, H., Contributions to the method of Lie series, Hochschultaschenbücher-Verlag, Stuttgart, 1967.

Hill, J.M., One dimensional Stefan problems: An introduction, Pitman Monographs and Surveys in Pure and Applied Mathematics 31, Longman, London, 1987.

Murphy, G.M., Ordinary differential equations and their solutions, Van Nostrand, New York, 1960.

Ovsjannikov, L.V., Group properties of differential equations, translation by G. Bluman, 1967.

Page, J.M., Ordinary differential equations with an introduction to Lie's theory of the group of one-parameter, Macmillian, New York, 1897.

ANSWERS AND HINTS

The following provides answers and hints to some of the problems at the end of each chapter. Roughly speaking there are more details for those problems dealing with basic issues and which the student must grasp before moving on. Those problems for which less information is provided are either already sufficiently structured in the text by a sequence of easier stages or the problem relates to issues which have already been adequately expounded either in the text as an example or as a prior problem. In addition solutions are not given for those problems which are purposely included as summaries of results given elsewhere. Such problems are generally harder and the majority of students will need to undertake further reading.

CHAPTER ONE

1. (i) $\beta = -\alpha/2$, take $\alpha = -2$ and $\beta = 1$; $x_1 = e^{-2\epsilon}x$, $y_1 = e^{\epsilon}y$,
   so that $u(x_1, y_1) = y_1^2 x_1 = y^2 x = u(x, y)$ and $\dfrac{du}{dx} = 2xy\dfrac{dy}{dx} + y^2 = 2xy\left(\dfrac{A}{x^{\frac{3}{2}}} + By^3\right) + y^2$,
   which gives $\dfrac{du}{dx} = \dfrac{2(Au^{1/2} + Bu^2) + u}{x}$ which is separable.

   (ii) $4\alpha = 3\beta$, take $\alpha = 3$ and $\beta = 4$; $x_1 = e^{3\epsilon}x$, $y_1 = e^{4\epsilon}y$,
   so that $u(x_1, y_1) = y_1^3 x_1^{-4} = y^3 x^{-4} = u(x, y)$ and $\dfrac{du}{dx} = \dfrac{3y^2}{x^4}\dfrac{dy}{dx} - \dfrac{4y^3}{x^5} = -5\dfrac{u}{x}\left(\dfrac{2-u}{1-2u}\right)$,
   which is separable.

   (iii) $\alpha = -n\beta$, take $\alpha = -n$ and $\beta = 1$; $x_1 = e^{-n\epsilon}x$, $y_1 = e^{\epsilon}y$,
   so that $u(x_1, y_1) = x_1 y_1^n = xy^n = u(x, y)$ and $\dfrac{du}{dx} = xny^{n-1}\dfrac{dy}{dx} + y^n = \dfrac{u}{x}\left(\dfrac{A - B + u}{A + u}\right)$,
   and again separable.

2. $\epsilon = 0$ and $-\epsilon$ characterize the identity and inverse respectively. Further if $x_2 = x_1 + \delta$, $y_2 = e^{-2\delta}y_1$ then $x_2 = x + (\epsilon + \delta)$, $y_2 = e^{-2(\epsilon+\delta)}y$ so that product is characterized by $\epsilon + \delta$. Further invariant $u(x_1, y_1) = \log y_1 + 2x_1 = \log y + 2x = u(x, y)$ so that $\dfrac{du}{dx} = \dfrac{1}{y}\dfrac{dy}{dx} + 2 = \dfrac{2u}{(u-1)}$ which can be readily integrated.

3. Invariant $u(x_1, y_1) = y_1 - x_1 = y - x = u(x, y)$ so that $\dfrac{du}{dx} = \dfrac{dy}{dx} - 1 = \dfrac{A^2}{u^2} - 1$ which can be integrated as follows, with $u = A\sin\theta$
   $$dx = \dfrac{u^2 du}{(A^2 - u^2)} = A\dfrac{\sin^2\theta}{\cos\theta}d\theta = A\dfrac{\tan^2\theta}{\sec\theta}d\theta = A(\cos\theta + \sec\theta)d\theta.$$

4. $\dfrac{dy}{dx} = \dfrac{\rho'(x-x_0)}{\rho(x)} - \dfrac{\rho(x-x_0)}{\rho(x)^2}\rho'(x) = \left(\dfrac{\rho(x-x_0)}{\rho(x)}\right)^2 - \dfrac{\rho(x-2x_0)}{\rho(x)}$ by (1.10),

$= y(x)^2 - y(x)y(x-x_0).$

5. $\dfrac{e^x}{f(t)} - \dfrac{e^x}{f(t)^2}f'(t)t = \dfrac{e^x}{f(t)} - \dfrac{e^x}{f(t)}\dfrac{t}{f(\lambda t)}$ by substitution into (1.11).

6. $y = x\omega$ and (1.13) give $x\omega\left(x\dfrac{d\omega}{dx} + \omega\right) = \dfrac{3x(1-x)}{(1+x)} + 2x\omega + \dfrac{(5+3x)x\omega^2}{(1+x)4}$.

$\dfrac{ds}{dx} = -\dfrac{2}{(1+x)^2};\ x\omega\dfrac{d\omega}{dx} = x\omega\dfrac{d\omega}{ds}\dfrac{ds}{dx} = -\left(\dfrac{1-s}{1+s}\right)\dfrac{\omega}{2}\dfrac{d\omega}{ds}(1+s)^2.$

7. $\xi = x/t^{1/2},\ c(x,t) = \phi(\xi),\ \dfrac{\partial c}{\partial t} = -\dfrac{\xi\phi'(\xi)}{2t},\ \dfrac{\partial^2 c}{\partial x^2} = \dfrac{\phi''(\xi)}{t}$;

$\phi''(\xi) + \dfrac{\xi}{2}\phi'(\xi) = 0;\ \dfrac{\phi''(\xi)}{\phi'(\xi)} + \dfrac{\xi}{2} = 0;\ \log\phi'(\xi) + \dfrac{\xi^2}{4} = \log A$;

$\phi'(\xi) = Ae^{-\xi^2/4}.$

8. Similar to previous problem.

9. $D(c) = c$ gives $\phi(\xi)\phi''(\xi) + \phi'(\xi)^2 + \xi\phi'(\xi)/2 = 0$;

$\phi = \xi^2\psi$ where ψ is an invariant of group.

$\xi^2\psi(\xi^2\psi'' + 4\xi\psi' + 2\psi) + (\xi^2\psi' + 2\xi\psi)^2 + \xi(\xi^2\psi' + 2\xi\psi)/2 = 0.$

Divide by ξ^2 to give equation of Euler type, use

$$\xi\psi' = \dfrac{d\psi}{dy},\ \xi^2\psi'' = \dfrac{d^2\psi}{dy^2} - \dfrac{d\psi}{dy},\ \dfrac{d^2\psi}{dy^2} = p\dfrac{dp}{d\psi}.$$

$p = -2\psi$ becomes $\xi\psi' + 2\psi = 0$ or $\xi^2\psi = $ constant.

10. $q = p\psi$ and $q\dfrac{dq}{d\psi} = p\psi\left(\psi\dfrac{dp}{d\psi} + p\right) = \psi\left(\psi p\dfrac{dp}{d\psi} + p^2\right).$

Now use differential equation for p.

Answers and Hints

11. $\dfrac{\partial z}{\partial t} = Be^{Bc}\dfrac{\partial c}{\partial t} = zAB\dfrac{\partial}{\partial x}\left(\dfrac{1}{B}\dfrac{\partial z}{\partial x}\right) = Az\dfrac{\partial^2 z}{\partial x^2}.$

12. $\dfrac{\partial z}{\partial t} = t^{(m-n)/n}\{m\phi(\xi) - \xi\phi'(\xi)\}/n,\quad \dfrac{\partial^2 z}{\partial x^2} = t^{(m-2)/n}\phi''(\xi).$

    Desired result follows noting that $t^{(m-n)/n} = t^{2(m-1)/n}$ so long as $m+n=2$.

    $\phi = \xi^2\psi$ gives $nA\xi^2\psi(\xi^2\psi'' + 4\xi\psi' + 2\psi) + \xi(\xi^2\psi' + 2\xi\psi) - m\xi^2\psi = 0,$

    or $nA\psi\left(\dfrac{d^2\psi}{dy^2} + 3\dfrac{d\psi}{dy} + 2\psi\right) + \left(\dfrac{d\psi}{dy} + 2\psi\right) - m\psi = 0.$

13. $c(x,t) = f(x) + g(t),\ \dfrac{\partial c}{\partial t} = g'(t),\ \dfrac{\partial c}{\partial x} = f'(x).$

    $e^{-Bg(t)}g'(t) = A(e^{Bf(x)}f'(x))' = \text{constant} = \lambda$ (say).

    $e^{-Bg(t)} = B\lambda(t_0 - t);\ e^{Bf(x)}f'(x) = \dfrac{\lambda}{A}(x - x_0);\ e^{Bf(x)} = \dfrac{B\lambda}{2A}[(x-x_0)^2 + C].$

    Hence $e^{Bc} = \left\{\dfrac{B\lambda}{2A}[(x-x_0)^2 + C]\right\}\dfrac{1}{B\lambda(t_0 - t)} = \left\{\dfrac{(x-x_0)^2 + C}{2A(t_0 - t)}\right\}.$

14. (i) Use fact that $-\epsilon$ characterizes inverse so that in particular $x = x_1 - \epsilon t_1,\ t = t_1$. Further

    $\dfrac{\partial c_1}{\partial t_1} = \dfrac{\partial c_1}{\partial x}\dfrac{\partial x}{\partial t_1} + \dfrac{\partial c_1}{\partial t}\dfrac{\partial t}{\partial t_1} = \dfrac{\partial c_1}{\partial t} - \epsilon\dfrac{\partial c_1}{\partial x},$

    $\dfrac{\partial c_1}{\partial x_1} = \dfrac{\partial c_1}{\partial x}\dfrac{\partial x}{\partial x_1} + \dfrac{\partial c_1}{\partial t}\dfrac{\partial t}{\partial x_1} = \dfrac{\partial c_1}{\partial x},$

    $\dfrac{\partial^2 c_1}{\partial x_1^2} = \dfrac{\partial}{\partial x}\left(\dfrac{\partial c_1}{\partial x_1}\right)\dfrac{\partial x}{\partial x_1} + \dfrac{\partial}{\partial t}\left(\dfrac{\partial c_1}{\partial x_1}\right)\dfrac{\partial t}{\partial x_1} = \dfrac{\partial}{\partial x}\left(\dfrac{\partial c_1}{\partial x_1}\right) = \dfrac{\partial^2 c_1}{\partial x^2}.$

    Now use $c_1 = ec$ where e denotes $\exp(-\epsilon x/2 - \epsilon^2 t/4)$ and evaluate

    $\dfrac{\partial c_1}{\partial t_1} - \dfrac{\partial^2 c_1}{\partial x_1^2} = \dfrac{\partial c_1}{\partial t} - \epsilon\dfrac{\partial c_1}{\partial x} - \dfrac{\partial^2 c_1}{\partial x^2}$

    $= e\left(\dfrac{\partial c}{\partial t} - \dfrac{\epsilon^2}{4}c\right) - \epsilon e\left(\dfrac{\partial c}{\partial x} - \dfrac{\epsilon}{2}c\right) - e\left(\dfrac{\partial^2 c}{\partial x^2} - \epsilon\dfrac{\partial c}{\partial x} + \dfrac{\epsilon^2}{4}c\right)$

    $= e\left(\dfrac{\partial c}{\partial t} - \dfrac{\partial^2 c}{\partial x^2}\right)$ as required.

    (ii) Follow the same steps as in part (i).

15. Proceed precisely as detailed in Example 1.3.

CHAPTER TWO

1. On differentiating with respect to ϵ we have

$$\frac{\partial(dr_1/d\epsilon, \theta_1)}{\partial(r,\theta)} + \frac{\partial(r_1, d\theta_1/d\epsilon)}{\partial(r,\theta)} = -\frac{r}{r_1^2}\frac{dr_1}{d\epsilon},$$

and now multiply by $\dfrac{\partial(r,\theta)}{\partial(r_1,\theta_1)} = \dfrac{r_1}{r}$ to deduce desired result. This equation can be written

$$\frac{\partial}{\partial r_1}\left(r_1\frac{dr_1}{d\epsilon}\right) + \frac{\partial}{\partial \theta_1}\left(r_1\frac{d\theta_1}{d\epsilon}\right) = 0,$$

from which there exists $\phi(r_1,\theta_1,\epsilon)$ with the required properties. If ϕ is independent of ϵ then $\dfrac{d}{d\epsilon}\phi(r_1,\theta_1) = \dfrac{\partial \phi}{\partial r_1}\dfrac{dr_1}{d\epsilon} + \dfrac{\partial \phi}{\partial \theta_1}\dfrac{d\theta_1}{d\epsilon} = 0$ so that ϕ is an invariant.

2. (i) $\dfrac{dr_1}{d\epsilon} = -\dfrac{(Ar_1^2 + B)}{r_1}$, $\dfrac{d\theta_1}{d\epsilon} = 2A\theta_1$, which on separately integrating and using $r_1 = r$ and $\theta_1 = \theta$ when $\epsilon = 0$ gives

$$\frac{\log(Ar_1^2 + B)}{2A} = -\epsilon + \frac{\log(Ar^2 + B)}{2A}, \quad \log \theta_1 = 2A\epsilon + \log \theta,$$

from which the given result follows.

(ii) $\dfrac{dr_1}{d\epsilon} = -Ar_1$, $\dfrac{d\theta_1}{d\epsilon} = 2A\theta_1 + 2B\log r_1 + B$, which on integrating the first equation gives $r_1 = e^{-A\epsilon}r$ so that the second becomes $\dfrac{d\theta_1}{d\epsilon} = 2A\theta_1 + 2B\log r + B(1 - 2A\epsilon)$ which we can re-write as $\dfrac{d}{d\epsilon}(e^{-2A\epsilon}\theta_1) = 2B\log r e^{-2A\epsilon} + \dfrac{d}{d\epsilon}(B\epsilon e^{-2A\epsilon})$ which readily integrates to yield $\theta_1 e^{-2A\epsilon} = B(\epsilon - (\log r)/A)e^{-2A\epsilon} +$ constant, from which the given result may be obtained.

Alternatively Use $r_1 = e^{-A\epsilon}r$ and $\phi(r_1,\theta_1) = \phi(r,\theta)$ to deduce expression for θ_1.

(iii) $\dfrac{dr_1}{d\epsilon} = -\dfrac{A}{4r_1}(1 + \cos^2 \theta_1 - \sin^2 \theta_1)$, $\dfrac{d\theta_1}{d\epsilon} = 0$, from which we have immediately $\theta_1 = \theta$ and therefore $\dfrac{r_1^2}{2} = -\dfrac{A}{2}\epsilon\cos^2\theta + \dfrac{r^2}{2}$ providing the required result.

Answers and Hints

3. Notice $x = (e^{\epsilon x_1} - 1)/\epsilon$, $y = e^{-\epsilon x_1} y_1$. Further

$$\frac{dx_1}{d\epsilon} = \frac{x}{\epsilon(1+\epsilon x)} - \frac{\log(1+\epsilon x)}{\epsilon^2} = \frac{(1 - \epsilon x_1 - e^{-\epsilon x_1})}{\epsilon^2},$$

$$\frac{dy_1}{d\epsilon} = xy = \frac{(1 - e^{-\epsilon x_1})}{\epsilon} y_1,$$

and since ϵ occurs explicitly on the right-hand sides of these equations, the system is non-autonomous and therefore does not generate a one-parameter group of transformations. In addition $-\epsilon$ does not characterize the inverse.

4. (ii) $L(\phi, \psi) = \xi(\phi, \psi)_x + \eta(\phi, \psi)_y$
$= \xi(\phi_x, \psi) + \xi(\phi, \psi_x) + \eta(\phi_y, \psi) + \eta(\phi, \psi_y)$
$= (\xi\phi_x + \eta\phi_y, \psi) + (\phi, \xi\psi_x + \eta\psi_y)$
$-\phi_x(\xi, \psi) - \phi_y(\eta, \psi) - \psi_x(\phi, \xi) - \psi_y(\phi, \eta)$
$= (L(\phi), \psi) + (\phi, L(\psi)) - \phi_x(\xi_x\psi_y - \xi_y\psi_x)$
$-\phi_y(\eta_x\psi_y - \eta_y\psi_x) - \psi_x(\phi_x\xi_y - \phi_y\xi_x) - \psi_y(\phi_x\eta_y - \phi_y\eta_x)$
$= (L(\phi), \psi) + (\phi, L(\psi)) - (\xi_x + \eta_y)(\phi_x\psi_y - \phi_y\psi_x)$
$= (L(\phi), \psi) + (\phi, L(\psi)) - \omega(\phi, \psi),$

from which the desired result follows.

5. (ii) $P(\chi_{n-1}) = \sum_{k=0}^{n-1} P(\phi_k, \psi_{n-1-k})$

$= \sum_{k=0}^{n-1}\{(L(\phi_k), \psi_{n-1-k}) + (\phi_k, L(\psi_{n-1-k}))\}$ by 4(ii),

$= \sum_{k=0}^{n-1}\{((k+1)\phi_{k+1}, \psi_{n-1-k}) + (\phi_k, (n-k)\psi_{n-k})\}$ by 4(i),

$= \sum_{j=1}^{n}\{(j\phi_j, \psi_{n-j}) + (\phi_j, (n-j)\psi_{n-j})\} + n(\phi_0, \psi_n),$

from $j = k+1$ in the first sum and simply changing k to j in the second summation. Hence

$$P(\chi_{n-1}) = \sum_{j=0}^{n} n(\phi_j, \psi_{n-j}) = n\chi_n,$$

which yields the desired result.

6. (i) Set $\rho = \epsilon\omega + \dfrac{\epsilon^2}{2!}P(\omega) + \dfrac{\epsilon^3}{3!}P^2(\omega) + \dfrac{\epsilon^4}{4!}P^3(\omega) + \ldots$

then using the first few terms of $\log(1+\rho) = \sum_{j=1}^{\infty} \dfrac{(-1)^{j-1}}{j}\rho^j$ we have

$\log(x_1, y_1) = \log(1+\rho)$

$= \rho - \dfrac{\rho^2}{2} + \dfrac{\rho^3}{3} - \dfrac{\rho^4}{4} + \ldots$

and this gives first few terms of desired result assuming $\rho^2 < 1$.

(ii) Similarly use the first few terms of $(1+\rho)^{-1} = \sum_{j=0}^{\infty}(-1)^j\rho^j$

so that we have $(x_1, y_1)^{-1} = (1+\rho)^{-1}$
$= 1 - \rho + \rho^2 - \rho^3 + \rho^4 = \ldots$

which also gives first few terms of given result assuming $|\rho| < 1$.

7. Details similar to Problems 4.

8. Details similar to Problem 5.

CHAPTER THREE

1. From $\dfrac{dy_1}{dx_1} + p(x_1)y_1 = q(x_1)y_1^n$, $x_1 = f(x)$, $y_1 = g(x)y$,

we have $\left\{\dfrac{g(x)dy + g'(x)ydx}{f'(x)dx}\right\} + p(f)g(x)y = q(f)g(x)^n y^n$,

which becomes $\dfrac{dy}{dx} + \left\{\dfrac{g'(x)}{g(x)} + p(f)f'(x)\right\}y = q(f)g(x)^{n-1}f'(x)y^n$, from which we may deduce

$$p(x) = \dfrac{g'(x)}{g(x)} + p(f)f'(x), \quad q(x) = q(f)g(x)^{n-1}f'(x),$$

so that exactly as in Section 3.2 $\eta + \xi p = C_1$, while in place of $(3.7)_2$ we find

$$\xi' + \xi\left\{\dfrac{q'}{q} - (n-1)p\right\} = (1-n)C_1,$$

for which the integrating factor is $q(x)/s(x)^{n-1}$ and the given result readily follows.

Answers and Hints

2. To find $u(x,y)$ (see (2.8)) we solve

$$\frac{dy_1}{dx_1} = \frac{\eta(x_1)y_1}{\xi(x_1)} = \frac{(C_1 - p(x_1)\xi(x_1))y_1}{\xi(x_1)} \text{ which becomes}$$

$$\frac{dy_1}{y_1} = \frac{C_1 q(x_1) s(x_1)^{1-n} dx_1}{\left\{(1-n)C_1 \int_{x_0}^{x_1} s(t)^{1-n} q(t) dt + C_2\right\}} - p(x_1) dx_1,$$

and on integration we have

$$\log y_1 = \frac{1}{(1-n)} \log\left\{(1-n)C_1 \int_{x_0}^{x_1} s(t)^{1-n} q(t) dt + C_2\right\} - \int_{x_0}^{x_1} p(t) dt = \text{constant},$$

and the given expression for $u(x,y)$ follows immediately from this equation.

To find $v(x,y)$ (see (2.9)) we solve $\frac{dx_1}{d\epsilon} = \xi(x_1)$ or $\frac{dx_1}{\xi(x_1)} = d\epsilon$ which becomes

$$\frac{q(x_1) s(x_1)^{1-n} dx_1}{\left\{(1-n)C_1 \int_{x_0}^{x_1} s(t)^{1-n} q(t) dt + C_2\right\}} = d\epsilon,$$

and on integration gives on using $x_1 = x$ when ϵ is zero

$$\frac{1}{(1-n)C_1} \log\left\{(1-n)C_1 \int_{x_0}^{x_1} s(t)^{1-n} q(t) dt + C_2\right\}$$

$$= \epsilon + \frac{1}{(1-n)C_1} \log\left\{(1-n)C_1 \int_{x_0}^{x} s(t)^{1-n} q(t) dt + C_2\right\},$$

assuming that C_1 is non-zero. To deduce the differential equation in canonical coordinates (u,v) we proceed as follows,

$$\frac{du}{dv} = \frac{du/dx}{dv/dx} = \frac{\frac{s(x)(dy/dx + p(x)y)}{\Delta^{1/(1-n)}} - \frac{s(x)yC_1 s(x)^{1-n} q(x)}{\Delta^{1/(1-n)+1}}}{\frac{s(x)^{1-n} q(x)}{\Delta}},$$

where for convenience Δ denotes $\left\{(1-n)C_1 \int_{x_0}^{x} s(t)^{1-n} q(t) dt + C_2\right\}$.

On using the given differential equation we have

$$\frac{du}{dv} = \frac{q(x)y^{n-1}u - C_1 \frac{uq(x)s(x)^{1-n}}{\Delta}}{\frac{s(x)^{1-n} q(x)}{\Delta}},$$

and this becomes $\frac{du}{dv} = u(s(x)y)^{n-1}\Delta - C_1 u$ so that

$$\frac{du}{dv} = u\left\{\frac{\Delta}{(s(x)y)^{1-n}} - C_1\right\} = u\left(\frac{1}{u^{1-n}} - C_1\right) = u(u^{n-1} - C_1).$$

$$dv = \frac{du}{u(u^{n-1} - C_1)} = \frac{1}{C_1}\left\{\frac{u^{n-2}}{(u^{n-1} - C_1)} - \frac{1}{u}\right\} du \text{ so that on integration}$$

$$C_1 v = \frac{\log(u^{n-1} - C_1)}{(n-1)} - \log u + \text{ constant, from which we obtain}$$

$(n-1)C_1 v = \log(1 - C_1 u^{1-n}) +$ constant and the given results follows.

3. Proceed as in Problem 2 but details are a good deal simpler.

4. From $\dfrac{dy_1}{dx_1} + p(x_1)y_1 = q(x_1) + r(x_1)y_1^2$ we may deduce

$$p(x) = \frac{g'(x)}{g(x)} + p(f)f'(x), \quad q(x) = q(f)\frac{f'(x)}{g(x)}, \quad r(x) = r(f)g(x)f'(x).$$

As in Section 3.2 we deduce (3.7) and the additional equation

$$\xi r' + r\xi' + r\eta = 0,$$

so that on eliminating η through $(3.7)_1$ we have

$$\xi' + \xi\left(\frac{r'}{r} - p\right) = -C_1 \text{ and } \xi' + \xi\left(\frac{q'}{q} + p\right) = C_1.$$

By addition we may obtain $2\xi' + \xi\left(\dfrac{q'}{q} + \dfrac{r'}{r}\right) = 0$ or $\xi^2 qr =$ constant and by subtraction we have $\xi\left(\dfrac{q'}{q} - \dfrac{r'}{r} + 2p\right) = 2C_1$ and these expressions are only consistent if C_1 is zero and $\dfrac{q'}{q} - \dfrac{r'}{r} + 2p = 0$ which yields the given condition.

5. Similar to Problem 4.

6. Similar to Problem 4.

7. Similar to Problem 4.

8. (i) $\dfrac{dy}{dx} = e\left\{\dfrac{dy^*}{dx} - \dfrac{a(x)}{2}y^*\right\},$

$\dfrac{d^2 y}{dx^2} = e\left\{\dfrac{d^2 y^*}{dx^2} - a(x)\dfrac{dy^*}{dx} - \left[\dfrac{1}{2}\dfrac{da}{dx} - \dfrac{a(x)^2}{4}\right]y^*\right\},$

where e denotes $\exp\left(-\dfrac{1}{2}\displaystyle\int_{x_0}^{x} a(t)dt\right)$. Substitution of these expressions into the given equation yields the desired result.

Answers and Hints 177

(ii) $\dfrac{dy}{dx} = e\dfrac{dy}{dx^*}$, $\dfrac{d^2y}{dx^2} = e^2\dfrac{d^2y}{dx^{*2}} - a(x)e\dfrac{dy}{dx^*}$,

where here e denotes $\exp\left(-\int_{x_0}^{x} a(t)dt\right)$ and equals $\dfrac{dx^*}{dx}$. Desired result follows immediately from these formulae.

9. Straightforward illustration of Problem 8.

10. From $\dfrac{d^2y_1}{dx_1^2} + p(x_1)y_1 = q(x_1)$, equation is invariant if in addition to (3.16) we have

$q(x) = q(f)g^3$ which gives $q(x) = [q(x)+\epsilon\xi q'(x)](1+3\epsilon\eta)$ so that $\xi q' + 3\eta q = 0$ or $\dfrac{q'}{q} = -\dfrac{3}{2}\dfrac{\xi'}{\xi}$ using $(3.17)_1$. Hence desired result follows by integration. For the second part

$$\dfrac{du}{dv} = \dfrac{du/dx}{dv/dx} = \dfrac{\xi^{-1/2}dy/dx - \xi^{-3/2}\xi' y/2}{\xi^{-1}} = \xi^{1/2}\dfrac{dy}{dx} - \dfrac{\xi'}{2\xi^{1/2}}y,$$

so that $\dfrac{d^2u}{dv^2} = \dfrac{d}{dv}\left(\dfrac{du}{dv}\right) = \dfrac{d}{dx}\left(\dfrac{du}{dv}\right)\Big/\dfrac{dv}{dx}$

$$= \xi^2\left\{\xi^{1/2}\dfrac{d^2y}{dx^2} - \left(\dfrac{\xi''}{2\xi^{1/2}} - \dfrac{\xi'^2}{4\xi^{3/2}}\right)y\right\}$$

$$= \xi^{3/2}\dfrac{d^2y}{dx^2} - \dfrac{1}{4\xi^{1/2}}(2\xi\xi'' - \xi'^2)y,$$

and the differential equation becomes

$$\dfrac{d^2u}{dv^2} = \xi^{3/2}\{q_0\xi^{-3/2} - p(x)y\} - \dfrac{1}{4\xi^{1/2}}(2\xi\xi'' - \xi'^2)y,$$

namely $\dfrac{d^2u}{dv^2} + \left\{\dfrac{1}{4}(2\xi\xi'' - \xi'^2) + p\xi^2\right\}u = q_0$ and the given equation follows immediately as in Section 3.3.

21. Directly from $(3.27)_1$ and $(3.27)_2$ we have

$q - p'/2$

$$= \dfrac{g'''}{g} + \dfrac{3g'^3}{g^3} - 4\dfrac{g'g''}{g^2} + p(f)gg' + q(f)g^3$$

$$- \dfrac{1}{2}\left\{2\dfrac{g'''}{g} - 8\dfrac{g'g''}{g^2} + 6\dfrac{g'^3}{g^3} + 2p(f)gg' + \dfrac{dp}{df}(f)g^3\right\}, \text{ using (3.26)}$$

$$= \left\{g(f) - \dfrac{1}{2}\dfrac{dp}{df}(f)\right\}g^3.$$

For the second part, let Δ denote $q - p'/2$ then from $\Delta(f)g(x)^3 = \Delta(x)$ we may deduce $3\eta\Delta + \xi\Delta' = 0$ so that $\eta = \xi'$ gives $\xi^3\Delta = $ constant, as required.

CHAPTER FOUR

1. $\mu = (\xi M + \eta N)^{-1}$ is an integrating factor if

$$\left(\frac{M}{\xi M + \eta N}\right)_y = \left(\frac{N}{\xi M + \eta N}\right)_x,$$

that is if $(\xi M + \eta N)(M_y - N_x) = M(\xi M + \eta N)_y - N(\xi M + \eta N)_x$.

Now if we put $F = -M/N$ in equation (4.6) this equation becomes

$$\xi(MN_x - NM_x) + \eta(MN_y - NM_y) = \eta_x N^2 - (\eta_y - \xi_x)MN - \xi_y M^2,$$

which can be shown to be identical with the integrating factor condition.

2. μ is an integrating factor for both differential equations if

$$(\mu M)_y = (\mu N)_x \text{ and } (\mu N)_y = -(\mu M)_x$$

so that $N\mu_x - M\mu_y = \mu(M_y - N_x)$ and $M\mu_x + N\mu_y = -\mu(M_x + N_y)$.

Introduce $R = \sqrt{M^2 + N^2}$, $\Theta = \tan^{-1}(M/N)$ and use above equations to deduce expressions for μ_x and μ_y, namely

$$\mu_x = \mu[\Theta_y - (\log R)_x], \quad \mu_y = -\mu[\Theta_x + (\log R)_y],$$

and required equation follows by equating expressions for μ_{xy}.

5. Following Example 4.3 we need to solve (4.6) as a first order partial differential equation and therefore have to solve

$$\frac{dx}{d\tau} = \xi(x), \quad \frac{dy}{d\tau} = \eta(x)y + \zeta(x), \quad \frac{dF}{d\tau} = (\eta'(x)y + \zeta'(x)) + (\eta(x) - \xi'(x))F,$$

so that we require to integrate

$$\frac{dy}{dx} - \frac{\eta}{\xi} = \frac{\zeta}{\xi}, \quad \frac{dF}{dx} + \frac{(\xi' - \eta)}{\xi}F = \frac{(\eta'y + \zeta')}{\xi}.$$

Answers and Hints

These equations have integrating factors $s(x)$ and $\xi s(x)$ respectively where $s(x)$ is defined by (4.22). Thus we have

$$ys(x) - \int_{x_0}^{x} s(t)\frac{\zeta(t)}{\xi(t)}dt = \text{constant} = A \text{ (say)}$$

$$\frac{d}{dx}\{\xi sF\} = \eta'\left(A + \int_{x_0}^{x} s(t)\frac{\zeta(t)}{\xi(t)}dt\right) + \zeta's,$$

and by integration by parts we obtain

$$\xi sF = \eta\left(A + \int_{x_0}^{x} s(t)\frac{\zeta(t)}{\xi(t)}dt\right) - \int_{x_0}^{x} s(t)\frac{\zeta(t)}{\xi(t)}\eta(t)dt + \zeta s - \int_{x_0}^{x} \zeta(t)\frac{ds}{dt} + \text{constant}.$$

From the defintion (4.22) of $s(x)$ we see that the integrals cancel to give

$$F(x,y) = \frac{\eta(x)}{\xi(x)}y + \frac{\zeta(x)}{\xi(x)} + \frac{\Phi\left(ys(x) - \int_{x_0}^{x} s(t)\frac{\zeta(t)}{\xi(t)}dt\right)}{\xi(x)s(x)},$$

as the most general $F(x,y)$.

7. From (4.15)-(4.18) and the previous problem we have one integral $s(x)y = A$. In addition we have to integrate

$$y\frac{dW}{dy} + W = \frac{(n-1)y^{-n}}{\eta(x)}$$ which on using $y = A/s$ becomes

$$\frac{dW}{dx} + \frac{\eta(x)}{\xi(x)}W = \frac{(n-1)}{A^n}\frac{s(x)^n}{\xi(x)}$$ so that we have

$$\frac{W}{s} = \frac{(n-1)}{A^n}\int_{x_0}^{x} \frac{s(t)^{n-1}}{\xi(t)}dt + \frac{\Phi(A)}{A^n}.$$

But $W = \dfrac{w}{\xi(x)y^{n-1}} = \dfrac{(F - y\eta(x)/\xi(x))^{-1}}{\xi(x)y^{n-1}}$, from which we may deduce the desired result.

In canonical coordinates we have

$$\frac{du}{dv} = \frac{du/dx}{dv/dx} = \frac{s(x)\left(\frac{dy}{dx} - \frac{\eta(x)}{\xi(x)}y\right)}{\left\{\frac{1}{\xi(x)y^{n-1}} + \frac{(1-n)}{u^n}s(x)\left(\frac{dy}{dx} - \frac{\eta(x)}{\xi(x)}y\right)\int_{x_0}^{x}\frac{s(t)^{n-1}}{\xi(t)}dt\right\}}$$

$$= \left\{\frac{(1-n)}{u^n}\int_{x_0}^{x}\frac{s(t)^{n-1}}{\xi(t)}dt + \frac{1}{\xi(x)y^{n-1}s(x)}\left(\frac{dy}{dx} - \frac{\eta(x)}{\xi(x)}y\right)^{-1}\right\}^{-1}$$

$$= \left\{\frac{(1-n)}{u^n}\int_{x_0}^{x}\frac{s(t)^{n-1}}{\xi(t)}dt + \frac{1}{\xi(x)y^{n-1}s(x)}\frac{\xi(x)}{ys(x)^{n-1}}\left\{(n-1)\int_{x_0}^{x}\frac{s(t)^{n-1}}{\xi(t)}dt + \Phi(u)\right\}\right\}^{-1}$$

$$= \left\{\frac{\Phi(u)}{u^n}\right\}^{-1} = \frac{u^n}{\Phi(u)}, \text{ as required.}$$

8. Normal form by Problem 8(i) of Chapter 3.

9. In (4.6) take $F(x,y) = q(x) - p(x)y + r(x)y^2$ so that

$$\xi(q' - p'y + r'y^2) + (\eta y + \zeta)(2ry - p) = (\eta'y + \zeta') + (\eta - \xi')(q - py + ry^2),$$

and the three given equations result on equating coefficients of y^2, y and y^0. For the second part we use

$$\eta = -\left(\xi' + \frac{r'}{r}\xi\right), \quad \zeta = \frac{-\xi''}{2r} + \frac{\xi'}{2r}\left(p - \frac{r'}{r}\right) + \frac{\xi}{2r}\left(p' - \frac{r''}{r} + \frac{r'^2}{r^2}\right),$$

to eliminate η and ζ from the third equation and show that the resulting equation for $\xi(x)$ admits the first integral which is given.

CHAPTER FIVE

1. If $\rho(x)$ is non-zero then from (5.11) we have

$$(\rho y + \xi)(G'y + H') + G(\rho'y^2 + \eta y + \zeta)$$
$$= (\eta - 2\xi')(Gy + H) + (\rho'''y^2 + \eta''y + \zeta'').$$

Remarkably coefficients of y^2 identically balance. From the coefficient of y we find

$$\xi G' + \rho H' + G\eta = (\eta - 2\xi')G + \eta'',$$

and using the given expressions for G and H this can be rearranged to give

$$\{\rho^2(\xi'' + \eta') - 3\rho\rho'\xi' - 3(\rho\rho'' - \rho'^2)\xi\}' = 0.$$

On integration, dividing by ρ^2 and a further integration, we obtain the desired result.

2. If $\rho(x)$ is zero and $\eta = \xi'/2$ then we need to solve $(5.11)_2$ as a first order partial differential equation, thus

$$\xi \frac{\partial F}{\partial x} + \left(\frac{\xi'y}{2} + \zeta\right)\frac{\partial F}{\partial y} = \frac{-3\xi'}{2}F + \left(\frac{\xi'''}{2}y + \zeta''\right).$$

From $\dfrac{dx}{d\tau} = \xi, \; \dfrac{dy}{d\tau} = \left(\dfrac{\xi'y}{2} + \zeta\right), \; \dfrac{dF}{d\tau} = \dfrac{-3\xi'}{2}F + \left(\dfrac{\xi'''}{2}y + \zeta''\right)$, we have

$$\frac{dy}{dx} - \frac{\xi'y}{2\xi} = \frac{\zeta}{\xi}, \quad \frac{dF}{dx} + \frac{3\xi'}{2\xi}F = \left(\frac{\xi'''}{2\xi}y + \frac{\zeta''}{\xi}\right).$$

Answers and Hints

Integrating factor for the first equation is $\xi^{-1/2}$ and we readily deduce

$$\frac{y}{\xi^{1/2}} - \int_{x_0}^{x} \frac{\zeta(t)}{\xi(t)^{3/2}} dt = \text{constant} = A \text{ (say)},$$

while for the second the integrating factor is $\xi^{3/2}$ and we have

$$\frac{d}{dx}(\xi^{3/2}F) = \frac{\xi\xi'''}{2}\left(A + \int_{x_0}^{x} \frac{\zeta(t)}{\xi(t)^{3/2}} dt\right) + \xi^{1/2}\zeta'',$$

and on using integration by parts we may deduce

$$\xi^{3/2}F = \frac{1}{2}\left(\xi\xi'' - \frac{\xi'^2}{2}\right)\left(A + \int_{x_0}^{x} \frac{\zeta(t)}{\xi(t)^{3/2}} dt\right) - \frac{1}{2}\int_{x_0}^{x}\left(\xi(t)\xi''(t) - \frac{\xi'(t)^2}{2}\right)\frac{\zeta(t)}{\xi(t)^{3/2}} dt$$
$$+ \xi^{1/2}\zeta' - \int_{x_0}^{x} \zeta'(t)\frac{\xi'(t)}{2\xi(t)^{1/2}} dt + \text{constant}.$$

Remarkably two integrals combine to become

$$-\frac{1}{2}\int_{x_0}^{x} \frac{d}{dt}\left\{\zeta(t)\frac{\xi'(t)}{\xi(t)^{1/2}}\right\} dt = -\frac{\zeta(x)\xi'(x)}{2\xi(x)^{1/2}} + \text{constant},$$

and altogether we obtain

$$F = \frac{1}{2}\left(\frac{\xi'}{\xi^{1/2}}\right)'\frac{y}{\xi^{1/2}} + \frac{1}{\xi^{1/2}}\left(\frac{\zeta}{\xi^{1/2}}\right)' + B,$$

where B is a constant. The desired result follows immediately from $B = \Phi(A)$.

3. In this case $F = -p(x)y^2$ and in the terminology of Example 5.1 we have that $\rho(x)$ is zero, $2\eta' = \xi''$ and $(5.11)_2$ becomes

$$\xi p' y^2 + (\eta y + \zeta)2py = (\eta - 2\xi')py^2 - (\eta''y + \zeta''),$$

from which we may deduce

$$\zeta'' = 0, \quad \eta'' = -2p\zeta, \quad \xi p' + (\eta + 2\xi')p = 0,$$

arising from the coefficients of y^0, y and y^2 respectively. Integrating the first two equations clearly involves four arbitrary constants to determine $\zeta(x)$ and $\eta(x)$. On integrating $\xi'' = 2\eta'$ we have a further two arbitrary constants giving at most six altogether. Notice however $\xi(x)$ and $\eta(x)$ must be compatible in the sense that $\xi p' + (\eta + 2\xi')p = 0$ which might reduce the number of one-parameter groups. For the second part, the given equation is invariant under

$$x_1 = e^\epsilon x, \quad y_1 = e^{-(m+2)\epsilon} y,$$

so that $u = yx^{(m+2)}$ is an invariant of the group. Accordingly set $y = x^{-(m+2)}u$ so that

$$\frac{dy}{dx} = x^{-(m+2)}\frac{du}{dx} - (m+2)x^{-(m+3)}u,$$

$$\frac{d^2y}{dx^2} = x^{-(m+2)}\frac{d^2u}{dx^2} - 2(m+2)x^{-(m+3)}\frac{du}{dx} + (m+2)(m+3)x^{-(m+4)}u,$$

and the given differential equation becomes

$$x^2\frac{d^2u}{dx^2} - 2(m+2)x\frac{du}{dx} + (m+2)(m+3)u + \alpha u^2 = 0,$$

which is of the Euler type so with $t = \log x$ and

$$x\frac{du}{dx} = \frac{du}{dt}, \quad x^2\frac{d^2u}{dx^2} = \frac{d^2u}{dt^2} - \frac{du}{dt},$$

we have

$$\frac{d^2u}{dt^2} - (2m+5)\frac{du}{dt} + (m+2)(m+3)u + \alpha u^2 = 0.$$

Finally $p = \dfrac{du}{dt}$ and $p\dfrac{dp}{du} = \dfrac{d^2u}{dt^2}$ yields the Abel equation of the second kind,

$$p\frac{dp}{du} - (2m+5)p + (m+2)(m+3)u + \alpha u^2 = 0.$$

4. For the one-parameter group in Problem 5 of Chapter 4 we have

$$A = ys(x) - \int_{x_0}^{x} s(t)\frac{\zeta(t)}{\xi(t)}dt, \quad B = s(x)(\xi(x)z - \eta(x)y - \zeta(x)),$$

where $s(x)$ is defined by equation (4.22). Now the most general second order equation arises from

$$\frac{dB}{dA} = \frac{dB/dx}{dA/dx} = \Phi(A, B),$$

Answers and Hints

which can be shown to become

$$\frac{\left\{\xi^2 \frac{d^2y}{dx^2} + \xi(\xi' - 2\eta)\frac{dy}{dx} + (\eta^2 - \xi\eta')y + (\eta\zeta - \xi\zeta')\right\}}{\left\{\xi\frac{dy}{dx} - \eta y - \zeta\right\}} = \Phi(A, B),$$

or $\dfrac{d^2y}{dx^2} + \left(\dfrac{\xi'}{\xi} - \dfrac{2\eta}{\xi}\right)\dfrac{dy}{dx} + \left(\dfrac{\eta^2}{\xi^2} - \dfrac{\eta'}{\xi}\right)y + \left(\dfrac{\eta\zeta}{\xi^2} - \dfrac{\zeta'}{\xi}\right) = \dfrac{\Phi_1(A, B)}{s\xi^2},$

which is the desired result.

For the one-parameter group in Problem 7 of Chapter 4 we have

$$A = s(x)y, \quad B = \frac{1}{s(x)[\xi(x)z - \eta(x)y]y^{n-1}} - \frac{(n-1)}{s(x)^n y^n}\int_{x_0}^{x} \frac{s(t)^{n-1}}{\xi(t)}dt,$$

where formally $s(x)$ is precisely as defined in the first part of the problem. In the usual way the most general second order equation is obtained from

$$\frac{dB}{dA} = \frac{dB/dx}{dA/dx} = \Phi(A, B),$$

which we simplify as follows. Introduce the canonical coordinates $v(x, y)$ employed in Problem 7 of Chapter 4, namely

$$v(x, y) = \frac{1}{y^{n-1}s(x)^{n-1}}\int_{x_0}^{x}\frac{s(t)^{n-1}}{\xi(t)}dt,$$

then we have

$$\frac{dv}{dA} = \frac{\left\{\frac{1}{\xi y^{n-1}} + \frac{(1-n)}{A^n}s\left(\frac{dy}{dx} - \frac{\eta}{\xi}y\right)\int_{x_0}^{x}\frac{s(t)^{n-1}}{\xi(t)}dt\right\}}{s\left(\frac{dy}{dx} - \frac{\eta}{\xi}y\right)}$$

$$= \left\{\frac{(1-n)}{A}v + \frac{1}{sy^{n-1}(\xi z - \eta y)}\right\}$$

$$= \left\{\frac{(1-n)}{A}v + B + \frac{(n-1)}{A}v\right\}$$

$$= B.$$

Thus using $B = \dfrac{1}{s(\xi z - \eta y)y^{n-1}} - \dfrac{(n-1)}{A}v$ we have,

$$\frac{dB}{dA} = \frac{-\xi[\xi y'' + (\xi' - \eta)y' - \eta' y]}{s^2(\xi y' - \eta y)^3 y^{n-1}} + \frac{\eta}{s^2(\xi y' - \eta y)^2 y^{n-1}}$$

$$- \frac{(n-1)\xi y'}{s^2(\xi y' - \eta y)^2 y^n} - \frac{(n-1)}{A}B + \frac{1}{A^2}\left(\frac{A}{s(\xi y' - \eta y)y^{n-1}} - BA\right)$$

$$= \Phi(A, B),$$

which becomes

$$\frac{\xi[\xi y'' + (\xi' - \eta)y' - \eta' y]}{s^2(\xi y' - \eta y)^3 y^{n-1}} + \frac{(n-2)\xi y'}{s^2(\xi y' - \eta y)^2 y^n} = \Phi_1(A, B),$$

and this equation simplifies to give

$$\xi \frac{d^2 y}{dx^2} + (\xi' - \eta)\frac{dy}{dx} - \eta' y + \frac{(n-2)}{y}\left(\xi \frac{dy}{dx} - \eta y\right)\frac{dy}{dx} = \frac{\Phi_1(A, B)}{\xi} s^2 \left(\xi \frac{dy}{dx} - \eta y\right)^3 y^{n-1}.$$

Notice that if $n = 1$ on using

$$-\frac{1}{y}\left(\xi \frac{dy}{dx} - \eta y\right)\frac{dy}{dx} = \frac{-1}{\xi y}\left(\xi \frac{dy}{dx} - \eta y\right)^2 - \frac{\eta}{\xi}\left(\xi \frac{dy}{dx} - \eta y\right),$$

so that the most general differential equation now becomes

$$\xi \frac{d^2 y}{dx^2} + (\xi' - 2\eta)\frac{dy}{dx} + \left(\frac{\eta^2}{\xi} - \eta'\right) y = \frac{1}{s\xi}\left\{\frac{\Phi_1(A, B)}{B^3} + \frac{1}{AB^2}\right\},$$

which agrees with equation (5.20) of Example 5.5 (noting that B in that example is B^{-1} used here).

CHAPTER SIX

1. In this case we have from (6.18)

$$\xi(x, t) = \kappa, \quad \eta(x, t) = \alpha, \quad \zeta(x, t) = \lambda,$$

so we have to solve

$$\frac{dx_1}{d\epsilon} = \kappa, \quad \frac{dt_1}{d\epsilon} = \alpha, \quad \frac{dc_1}{d\epsilon} = \lambda c_1,$$

subject to

$$x_1 = x, \quad t_1 = t, \quad c_1 = c,$$

when $\epsilon = 0$ and the given result follows immediately. The functional form of the solution is obtained by solving the partial differential equation,

$$\kappa \frac{\partial c}{\partial x} + \alpha \frac{\partial c}{\partial t} = \lambda c,$$

Answers and Hints 185

and from the characteristic equations

$$\frac{dx}{d\tau} = \kappa, \quad \frac{dt}{d\tau} = \alpha, \quad \frac{dc}{d\tau} = \lambda c,$$

we may deduce

$$\frac{dx}{dt} = \frac{\kappa}{\alpha}, \quad \frac{dc}{dx} = \frac{\lambda}{\kappa}c,$$

and therefore

$$\alpha x - \kappa t = \text{constant}, \quad \log c = \frac{\lambda}{\kappa}x + \text{constant},$$

from which the given functional form follows. On substituting this into the diffusion equation (6.3), the following differential equation with constant coefficients is obtained,

$$(\alpha\kappa)^2 \phi'' + (2\alpha\lambda + \kappa^2)\kappa\phi' + \lambda^2 \phi = 0.$$

In the notation of Problem 14 of Chapter 5, this means that the coupled equations for $A(x)$ and $B(x)$ given there, have the special solution

$$A(x) = A_0 e^{-nx}, \quad B(x) = nx,$$

where A_0 and n are constants and $n = k/2^{1/2}$.

2. In this case we have from (6.18)

$$\xi(x,t) = \beta x + \delta t, \quad \eta(x,t) = 2\beta t, \quad \zeta(x,t) = -\delta x/2,$$

from which we may immediately deduce $t_1 = e^{2\beta\epsilon}t$, so that

$$\frac{dx_1}{d\epsilon} = \beta x_1 + \delta t_1 = \beta x_1 + \delta e^{2\beta\epsilon}t,$$

and therefore

$$\frac{d}{d\epsilon}(e^{-\beta\epsilon}x_1) = \delta e^{\beta\epsilon}t,$$

which on integration gives

$$e^{-\beta\epsilon}x_1 = \frac{\delta}{\beta}e^{\beta\epsilon}t + x - \frac{\delta}{\beta}t,$$

which gives the desired result. Further we have

$$\frac{dc_1}{d\epsilon} = -\frac{\delta x}{2}c_1 = -\frac{\delta c_1}{2}\left\{xe^{\beta\epsilon} + \frac{t\delta}{\beta}(e^{2\beta\epsilon} - e^{\beta\epsilon})\right\},$$

which integrates to give

$$\log c_1 = -\frac{\delta}{2}\left\{\frac{xe^{\beta\epsilon}}{\beta} + \frac{t\delta}{\beta^2}\left(\frac{e^{2\beta\epsilon}}{2} - e^{\beta\epsilon}\right)\right\} + \frac{\delta}{2}\left\{\frac{x}{\beta} + \frac{t\delta}{\beta^2}\left(\frac{1}{2} - 1\right)\right\} + \log c,$$

and this can be rearranged to yield the stated result. In order to find the functional form we need to deduce two integrals of

$$\frac{dx}{dt} = \frac{\beta x + \delta t}{2\beta t}, \quad \frac{dc}{dt} = \frac{-\delta x}{4\beta t}c.$$

First set $v = x/t$ so that $x = vt$ and

$$\frac{dx}{dt} = t\frac{dv}{dt} + v = \frac{v}{2} + \frac{\delta}{2\beta},$$

which becomes

$$\frac{d}{dt}(t^{1/2}v) = \frac{\delta}{2\beta t^{1/2}},$$

so that on integration we obtain

$$t^{1/2}v = \frac{\delta}{\beta}t^{1/2} + A,$$

where A denotes a constant. Thus the second equation gives

$$\frac{dc}{dt} = -\frac{\delta}{4\beta}\left(\frac{\delta}{\beta} + \frac{A}{t^{1/2}}\right)c,$$

which on integrating as a separable equation becomes

$$\log c = \frac{-\delta}{4\beta}\left(\frac{\delta}{\beta}t + 2At^{1/2}\right) + B,$$

where B denotes a second constant and the given functional form follows on taking B as an arbitrary function of A. With ω defined by

$$\omega = \delta t^{1/2} - \beta x t^{-1/2},$$

the differential equation for $\phi(\omega)$ can be shown to become

$$2\beta^2\phi''(\omega) + \omega\phi'(\omega) = 0.$$

Answers and Hints

3. From $\dfrac{\partial c}{\partial t} = \dot\psi(t)\phi(y) - 2y\dfrac{\dot X(t)}{X(t)}\psi(t)\phi'(y)$ and $\dfrac{\partial^2 c}{\partial x^2} = \dfrac{\psi(t)}{X(t)^2}\{4y\phi''(y) + 2\phi'(y)\}$,

we may deduce the given equation which is sensible provided $X(t)\dot X(t) = \beta$ where β is a constant. Further if a denotes the separation constant then we have

$$\frac{\dot\psi(t)}{\psi(t)} = \frac{4a}{(\alpha + 2\beta t)},$$

which on integration gives

$$\log \psi(t) = \frac{4a}{2\beta}\log(\alpha + 2\beta t) + \text{ constant,}$$

so that $\psi(t) = \psi_0 X(t)^{4a/\beta}$ as required where ψ_0 denotes an arbitrary constant. The given equation for $\phi(y)$ follows immediately and is changed to a confluent hypergeometric equation by setting $y = -2z/\beta$.

10. We have

$$\frac{\partial c}{\partial t} = \alpha(x)\frac{\partial C}{\partial t}, \quad \frac{\partial c}{\partial x} = \alpha(x)\beta'(x)\frac{\partial C}{\partial y} + \alpha'(x)C,$$

$$\frac{\partial^2 c}{\partial x^2} = \alpha(x)\beta'(x)^2\frac{\partial^2 C}{\partial y^2} + (2\alpha'(x)\beta'(x) + \alpha(x)\beta''(x))\frac{\partial C}{\partial y} + \alpha''(x)C,$$

and from the equation

$$\frac{\partial c}{\partial t} = p(x)\frac{\partial^2 c}{\partial x^2} + p'(x)\frac{\partial c}{\partial x},$$

we may deduce the three equations,

$$p(x)\alpha''(x) + p'(x)\alpha'(x) = 0,$$

$$p(x)(2\alpha'(x)\beta'(x) + \alpha(x)\beta''(x)) + p'(x)\alpha(x)\beta'(x) = 0,$$

and $p(x)\beta'(x)^2 = 1$. The first two integrate immediately to yield

$$p(x)\alpha'(x) = A, \quad p(x)\beta'(x)\alpha(x)^2 = B,$$

where A and B denote arbitrary constants. Thus

$$\alpha'(x) = A\beta'(x)^2, \quad \alpha(x)^2 = B\beta'(x),$$

so that $\alpha'(x) = A\alpha(x)^4/B^2$ which integrates to give

$$-\frac{1}{3\alpha(x)^3} = \frac{A}{B^2}x - \frac{C}{3},$$

where C is a further arbitrary constant. Hence we have

$$\alpha(x) = (C - 3Ax/B^2)^{-1/3}.$$

Finally, from

$$p(x) = \frac{A}{\alpha'(x)} = \frac{B^2}{\alpha(x)^4} = B^2\left(C - \frac{3Ax}{B^2}\right)^{4/3},$$

we may deduce the desired result with constants and C_1 and C_2 as follows,

$$C_1 = -3AB^{-1/2}, \quad C_2 = B^{3/2}C.$$

15. Firstly it follows that $\xi(x_0,0) = \eta(x_0,0) = 0$ so that the point $x = x_0$ at time $t = 0$ remains unchanged by the one-parameter group (6.1). Secondly the condition

$$\zeta(x_0,0) = -\frac{\partial \xi}{\partial x}(x_0,0),$$

arises from the initial condition,

$$\int_{-\infty}^{\infty} \phi(x)c(x,0)dx = c_0\phi(x_0),$$

which holds for every test function $\phi(x)$. Thus from

$$\int_{-\infty}^{\infty} \phi(x_1)c_1(x_1,0)dx_1 = c_0\phi(x_0),$$

we may deduce using (6.1)

$$\int_{-\infty}^{\infty} \phi\bigl(x + \epsilon\xi(x,0)\bigr)\bigl(1 + \epsilon\zeta(x,0)\bigr)\left(1 + \epsilon\frac{\partial \xi}{\partial x}(x,0)\right)c(x,0)dx = c_0\phi(x_0),$$

which becomes

$$\int_{-\infty}^{\infty} \bigl(\phi(x) + \epsilon\xi(x,0)\phi'(x)\bigr)\left(1 + \epsilon\bigl(\zeta(x,0) + \frac{\partial \xi}{\partial x}(x,0)\bigr)\right)c(x,0)dx = c_0\phi(x_0).$$

Answers and Hints

Thus noting that $c(x,0) = c_0 \delta(x - x_0)$ we observe that this equation yields

$$\left(\phi(x_0) + \epsilon \xi(x_0, 0)\phi'(x_0)\right)\left(1 + \epsilon(\zeta(x_0, 0) + \frac{\partial \xi}{\partial x}(x_0, 0))\right)c_0 = c_0 \phi(x_0),$$

and the desired condition follows on noting that $\xi(x_0, 0) = 0$. The given initial values of $\eta(t), \rho(t)$ and $\sigma(t)$ now readily follow from these conditions and (6.47) and (6.48).

16. For case(i) $\rho(t)$ is non-zero and $f(I)$ is given by (6.56). On solving (6.52) we find that solutions $\eta(t), \rho(t)$ and $\sigma(t)$ satisfying the conditions of the previous problem involve three arbitrary constants. Each constant corresponds to a one-parameter group leaving the boundary value problem unchanged and we may employ any one of these groups to determine $c(x,t)$. We therefore select the simplest which is given by

$$\eta(t) = 0, \quad \rho(t) = 2\sinh(\beta t), \quad \sigma(t) = \beta I_0 + (C_2/\beta)(1 - \cosh(\beta t)),$$

and this choice yields the given expressions for $\xi(x,t), \eta(x,t)$ and $\zeta(x,t)$. On solving (6.2) we find

$$c(x,t) = \frac{\phi(t)v(I)}{p(x)^{1/2}} \exp\left\{-\frac{\beta I^2}{4}\coth(\beta t) + \frac{[\beta^2 I_0 + C_2(1 - \cosh(\beta t))]I}{2\beta \sinh(\beta t)}\right\},$$

where $\phi(t)$ denotes an arbitrary function of t and in deriving this result we have made use of

$$J = -2\frac{d}{dI}\left\{\log\left(\frac{v(I)}{p(x)^{1/2}}\right)\right\},$$

which follows from $(6.45)_2$, (6.60) and (6.66). On writing (6.4) in the form

$$\frac{\partial c}{\partial t} = \frac{1}{p(x)^{1/2}}\frac{\partial}{\partial I}\left\{p(x)^{1/2}c\frac{\partial}{\partial I}\left[\log\left(\frac{p(x)^{1/2}c}{v(I)}\right)\right] - p(x)^{1/2}c\frac{v'(I)}{v(I)}\right\},$$

we may readily deduce a first order ordinary differential equation for $\phi(t)$ which can be solved to give the desired result. For the final part we need to assume $I(x) \to \pm\infty$ as $x \to \pm\infty$ so that we have

$$\lim_{t\to 0}\int_{-\infty}^{\infty} c(x,t)dx = \frac{2\phi_0 v(I_0)\pi^{1/2}}{\beta^{1/2}} = c_0,$$

which completely determines ϕ_0 and therefore $c(x,t)$.

17. For case(ii) $\rho(t)$ is zero, $f(I)$ is given by (6.57) and we have

$$\eta(0) = 0, \ \eta'(0) = 0, \ \sigma(0) = \frac{\eta''(0)}{8} I_0^2,$$

and from (6.54) we obtain

$$\eta(t) = \sinh^2(\beta t), \ \sigma(t) = \frac{(\beta I_0)^2}{4} - \frac{\beta}{4}\sinh(2\beta t) - \frac{C_3}{4}\sinh^2(\beta t).$$

Thus from (6.47), (6.48) and the partial differential equation (6.2) the given functional form follows on making use of the expression for J noted in the previous solution. Also using the form of the Fokker-Planck equation noted above and the functional form we can make use of (6.67) and (6.57) to deduce the desired result. The final part follows on using the asymptotic value for $I_n(\Omega)$, namely

$$I_n(\Omega) \sim \frac{e^\Omega}{(2\pi\Omega)^{1/2}} \text{ as } \Omega \to \infty,$$

and assuming $I(x) \to \pm\infty$ as $x \to \pm\infty$ so that Φ_0 has the value $\beta c_0 I_0^{1/2}/2v(I_0)$. To show that agreement is obtained with the result of the previous problem when $C_2 = 0$ and $C_4 = 0$ we need to use

$$I_{1/2}(z) = \left(\frac{2}{\pi z}\right)^{1/2} \sinh(z), \ I_{-1/2}(z) = \left(\frac{2}{\pi z}\right)^{1/2} \cosh(z).$$

18. (i) In this case we have $p(x) = 1, q(x) = bx$ and from (6.45) we obtain

$I(x) = x, J(x) = bx$, while (6.60) gives $u = -bx = -bI$. Therefore from (6.66) we have apart from a multiplicative constant, $v(I) = \exp(-bI^2/4)$ and from (6.59) we obtain $C_1 = b^2, C_2 = 0$ and $C_3 = -2b$. The given expression now follows from the results given in Problem 16.

(ii) In this case we have $p(x) = ax, q(x) = a + bx$ and from (6.45) we obtain

$$I(x) = 2\left(\frac{x}{a}\right)^{1/2}, J(x) = \frac{3}{2}\left(\frac{a}{x}\right)^{1/2} + b\left(\frac{x}{a}\right)^{1/2},$$

so that the function $\phi(x)$ defined by (6.50) becomes

$$\phi(x) = \frac{3a}{4x} + \frac{b^2 x}{a} = \frac{b^2}{4}I^2 + \frac{3}{I^2},$$

and therefore in the notation of case(ii) we have $C_1 = b^2/4, C_3 = 0$ and $C_4 = 3$. From (6.59), (6.60) and (6.66) we deduce that apart from a multiplicative constant $v(I)$ is given by

$$v(I) = I^{-1/2}\exp(-bI^2/8).$$

Answers and Hints 191

Now from the limit t tending to zero in Problem 17 we have

$$\int_{-\infty}^{\infty} c(x,t)dx = \frac{-2\Phi_0 v(I_0)}{b(\pi I_0 t a)^{1/2}} \int_{-\infty}^{\infty} \exp\left\{-\frac{(|x|^{1/2} - |x_0|^{1/2})^2}{at}\right\} \frac{dx}{|x|^{1/2}},$$

and the integral on the right hand side becomes

$$2\int_0^{\infty} \exp\left\{-\frac{(x^{1/2} - |x_0|^{1/2})^2}{at}\right\} \frac{dx}{x^{1/2}} = 4(at)^{1/2} \int_{-(|x_0|/at)^{1/2}}^{\infty} e^{-y^2} dy,$$

from which in the limit of t zero we may deduce

$$\lim_{t \to 0} \int_{-\infty}^{\infty} c(x,t)dx = \frac{-8\Phi_0 v(I_0)}{bI_0^{1/2}} = c_0.$$

The desired result now follows from the solution to Problem 17 noting that $n = 1$.

(iii) In this case we have $p(x) = x^2, q(x) = (b+2)x, I(x) = \log|x|, J(x) = (b+3)$ and $\phi(x) = (b+1)^2$ so that in the notation of case(i) $C_1 = C_2 = 0$ and $C_3 = (b+1)^2$. From $u = -(b+1)$ we have apart from a multiplicative constant

$$v(I) = \exp\{-(b+1)I/2\},$$

and the desired result follows from Problem 16.

20. (i) $\xi(x,t) = f(x+t) - g(x-t),$

$\eta(x,t) = f(x+t) + g(x-t),$

$\zeta(x,t,c) = \lambda c + F(x+t) + G(x-t),$

where f, g, F and G denote arbitrary functions and λ is a constant.

(ii) Observe that the transformation

$$c(x,t) = e^{-t/2\sigma} C(x,t),$$

reduces the telegrapher's equation to

$$\sigma \frac{\partial^2 C}{\partial t^2} = \frac{\partial^2 C}{\partial x^2} + \frac{C}{4\sigma},$$

which becomes the Klein-Gordon equation on setting $t^* = \sigma^{1/2} t$.

(iii) Observe that the transformations

$$x^* = x - \delta t, \quad t^* = t, \quad c(x,t) = C(x^*, t^*)$$

reduce the diffusion equation with convection to the classical diffusion equation

$$\frac{\partial C}{\partial t^*} = \frac{\partial^2 C}{\partial x^{*2}}.$$

(iv) $\xi(x,t) = \alpha x + \beta$,

$\eta(x,t) = \alpha t + \gamma$,

$\zeta(x,t,c) = \delta c + f(x,t)$,

where α, β, γ and δ are all constants and $f(x,t)$ is any solution of the Klein-Gordon equation.

(v) $\xi(x,t) = 9\alpha x^2 + 4\alpha t^3 + 6\beta x + \gamma$,

$\eta(x,t) = 12\alpha x t + 4\beta t$,

$\zeta(x,t,c) = -(3\alpha x + \delta)c + f(x,t)$,

where α, β, γ and δ are all constants and $f(x,t)$ is any solution of the Tricomi equation.

21. (i) $\xi(x,y) = u(x,y)$, $\eta(x,y) = v(x,y)$, $\zeta(x,y) = w(x,y)$,

where u, v and w all satisfy the Laplace equation and u and v are harmonic conjugates, namely

$$\frac{\partial u}{\partial x} = \frac{\partial v}{\partial y}, \quad \frac{\partial u}{\partial y} = -\frac{\partial v}{\partial x}.$$

(ii) $\xi(x,y) = \alpha y + \beta$,

$\eta(x,y) = -\alpha x + \gamma$,

$\zeta(x,y,c) = \delta c + f(x,y)$,

where α, β, γ and δ are all constants and $f(x,y)$ is any solution of the Helmholtz equation.

Answers and Hints 193

CHAPTER SEVEN

1. We need to solve the characteristic equations

$$\frac{dx}{d\tau} = x + \kappa, \quad \frac{dt}{d\tau} = 2(t+\delta), \quad \frac{dc}{d\tau} = 0,$$

so that

$$\frac{dx}{dt} = \frac{x+\kappa}{2(t+\delta)}, \quad \frac{dc}{dt} = 0,$$

and therefore

$$\frac{x+\kappa}{(t+\delta)^{1/2}} = \text{constant}, \quad c = \text{constant},$$

and the given functional form follows on taking one constant to be an arbitrary function of the other. From

$$\frac{\partial c}{\partial t} = -\frac{\omega \phi'(\omega)}{2(t+\delta)}, \quad \frac{\partial c}{\partial x} = \frac{\phi'(\omega)}{(t+\delta)^{1/2}},$$

and (7.3) the given ordinary differential equation can be readily obtained.

2. The functional form and resulting ordinary differential equation can be obtained in a similar manner as outlined for the previous problem. Further it is not difficult to check that the ordinary differential equation is indeed invariant under the given group, which has canonical coordinate $u = \phi \omega^{-2/m}$ so we let $\phi = \omega^{2/m} u$ to obtain

$$\phi' = \omega^{2/m} u' + \frac{2}{m}\omega^{2/m-1}u, \quad \phi'' = \omega^{2/m}u'' + \frac{4}{m}\omega^{2/m-1}u' + \frac{2}{m}\left(\frac{2}{m}-1\right)\omega^{2/m-2}u.$$

Thus we have

$$\omega^2 u^m \left\{ \omega^{2/m} u'' + \frac{4}{m}\omega^{2/m-1}u' + \frac{2}{m}\left(\frac{2}{m}-1\right)\omega^{2/m-2}u \right\}$$

$$+ m\omega^{2(m-1)/m} u^{m-1} \left\{ \omega^{4/m} u'^2 + \frac{4}{m}\omega^{4/m-1}uu' + \frac{4}{m^2}\omega^{4/m-2}u^2 \right\}$$

$$+ \frac{(1+\lambda)}{2\alpha}\omega \left\{ \omega^{2/m}u' + \frac{2}{m}\omega^{2/m-1}u \right\} - \frac{\lambda}{\alpha m}\omega^{2/m}u = 0,$$

and on cancelling $\omega^{2/m}$ we may deduce

$$u^m \left\{ \omega^2 u'' + \frac{4}{m}\omega u' + \frac{2}{m}\left(\frac{2}{m} - 1\right)u \right\}$$

$$+ mu^{m-1}\left\{ \omega^2 u'^2 + \frac{4}{m}\omega u u' + \frac{4}{m^2}u^2 \right\}$$

$$+ \frac{(1+\lambda)}{2\alpha}\left\{ \omega u' + \frac{2}{m}u \right\} - \frac{\lambda u}{\alpha m} = 0.$$

This is an equation of the Euler type so we set $v = \log \omega$ and use

$$\omega u' = \frac{du}{dv}, \quad \omega^2 u'' + \omega u' = \frac{d^2 u}{dv^2},$$

to yield

$$u^m \left(\frac{d^2 u}{dv^2} - \frac{du}{dv}\right) + 4u^m \left(\frac{1}{m} + 1\right)\frac{du}{dv} + \frac{2}{m}\left(\frac{2}{m} + 1\right)u^{m+1}$$

$$+ mu^{m-1}\left(\frac{du}{dv}\right)^2 + \frac{(1+\lambda)}{2\alpha}\frac{du}{dv} + \frac{u}{\alpha m} = 0.$$

Hence with $p = du/dv$ we can finally deduce the first order ordinary differential equation

$$u^m p \frac{dp}{du} + mu^{m-1}p^2 + \left\{ \left(\frac{4}{m} + 3\right)u^m + \frac{(1+\lambda)}{2\alpha} \right\}p + \frac{2}{m}\left(\frac{2}{m} + 1\right)u^{m+1} + \frac{u}{\alpha m} = 0,$$

from which it is again evident that the exponents $m = -2$ and $m = -4/3$ play preferred roles.

3. We have to deduce two integrals for

$$\frac{dx}{dt} = \frac{\mu x^2 + (\lambda + 1)x + \kappa}{2(t + \delta)} = \frac{[2\mu x + (\lambda + 1)]^2}{8\mu(t + \delta)},$$

$$\frac{dc}{dx} = \frac{-3(c + \beta)(2\mu x + \lambda)}{2[\mu x^2 + (\lambda + 1)x + \kappa]} = \frac{-6\mu(c + \beta)(2\mu x + \lambda)}{[2\mu x + (\lambda + 1)]^2},$$

assuming $(\lambda + 1)^2 = 4\mu\kappa$. On integrating these as separable differential equations we obtain

$$-\frac{1}{[2\mu x + (\lambda + 1)]} = \frac{1}{4}\log(t + \delta) + \text{ constant},$$

$$\log(c + \beta) = -3\left\{\frac{1}{[2\mu x + (\lambda + 1)]} + \log[2\mu x + (\lambda + 1)]\right\} + \text{ constant},$$

Answers and Hints 195

from which the given functional form may be readily deduced. In the following we use the abbreviations,

$$e = \exp\left\{\frac{1}{(2\mu x + \lambda + 1)}\right\}, \quad b = 2\mu x + (\lambda + 1).$$

We have,

$$\frac{\partial c}{\partial t} = -\frac{\omega \phi'(\omega)(2\mu)^3}{2(t+\delta)(be)^3} = -\frac{4\mu^3}{b^3}\omega^3 e\phi'(\omega),$$

$$\alpha(c+\beta)^{-4/3}\frac{\partial c}{\partial x} = -3\alpha\frac{\partial}{\partial x}(c+\beta)^{-1/3}$$

$$= -\frac{3\alpha}{2\mu}\frac{\partial}{\partial x}\{\phi^{-1/3}be\}$$

$$= -\frac{3\alpha}{2\mu}\left\{-\frac{be}{3}\phi^{-4/3}\phi'\frac{4\mu\omega}{b^2} + 2\mu\phi^{-1/3}e - \phi^{-1/3}be\frac{2\mu}{b^2}\right\}$$

$$= 3\alpha e\left\{\frac{\phi^{-1/3}}{b} - \phi^{-1/3} + \frac{2}{3}\phi^{-4/3}\phi'\frac{\omega}{b}\right\},$$

and from this we may deduce

$$\frac{\partial}{\partial x}\left\{\alpha(c+\beta)^{-4/3}\frac{\partial c}{\partial x}\right\} = \frac{6\alpha\mu e}{b^3}\left\{\frac{4}{3}\omega(\phi^{-4/3}\phi'\omega)' - \phi^{-1/3} - \frac{4}{3}\omega\phi^{-4/3}\phi'\right\},$$

so that we have

$$\frac{4}{3}\phi^{-4/3}\omega^2\phi'' - \frac{16}{9}\phi^{-7/3}\omega^2\phi'^2 - \phi^{-1/3} + \frac{2\mu^2}{3\alpha}\omega^3\phi' = 0,$$

and the desired equation now follows. Let $\phi = \omega^{-3/2}u$ then we have

$$\phi' = \omega^{-3/2}u' - \frac{3}{2}\omega^{-5/2}u, \quad \phi'' = \omega^{-3/2}u'' - 3\omega^{-5/2}u' + \frac{15}{4}\omega^{-7/2}u,$$

and therefore

$$\omega^2 u'' - 3\omega u' + \frac{15}{4}u - \frac{4}{3u}\left(\omega^2 u'^2 - 3\omega u u' + \frac{9}{4}u^2\right) - \frac{3}{4}u + \frac{\mu^2}{2\alpha}\left(\omega u' - \frac{3}{2}u\right)u^{4/3} = 0.$$

In the usual way with $v = \log \omega$ and $p = du/dv$ we obtain

$$p\frac{dp}{du} - \frac{4}{3u}p^2 + \frac{\mu^2}{2\alpha}\left(p - \frac{3u}{2}\right)u^{4/3} = 0.$$

11. Invariance under the given group and the functional form can be deduced in a routine manner. From the functional form and Burger's equation we obtain

$$D\phi'' = \phi\phi' - (\omega\phi)'/2,$$

which can be readily integrated to give

$$D\phi' = \frac{\phi^2}{2} - \frac{\omega\phi}{2} + \text{constant},$$

but the integration constant is zero because ϕ and ϕ' vanish at infinity. Thus we may deduce

$$\left(\frac{1}{\phi}\right)' - \frac{\omega}{2D}\left(\frac{1}{\phi}\right) = \frac{-1}{2D},$$

and therefore

$$\frac{e^{-\omega^2/4D}}{\phi} = -\frac{(C + \int_{-\infty}^{\omega} e^{-\lambda^2/4D}\,d\lambda)}{2D},$$

where C is the constant of integration. Thus the given expression for $\phi(\omega)$ now follows. Now we require

$$\int_{-\infty}^{\infty} u(x,t)\,dx = \int_{-\infty}^{\infty} \phi(\omega)\,d\omega = u_0,$$

so from the given expression for $\phi(\omega)$ we obtain

$$\left[\log(C + \int_{-\infty}^{\omega} e^{-\lambda^2/4D}\,d\lambda)\right]_{-\infty}^{\infty} = -\frac{u_0}{2D},$$

and using

$$\int_{-\infty}^{\infty} e^{-\lambda^2/4D}\,d\lambda = 2D^{1/2}\int_{-\infty}^{\infty} e^{-x^2}\,dx = 2(\pi D)^{1/2},$$

we obtain

$$1 + \frac{2(\pi D)^{1/2}}{C} = e^{-u_0/2D},$$

or

$$C = \frac{2(\pi D)^{1/2}}{e^{-u_0/2D} - 1} = \frac{-2(\pi D)^{1/2}e^{u_0/4D}}{(e^{u_0/4D} - e^{-u_0/4D})},$$

from which the desired result can be readily deduced.

Answers and Hints

13. From $c(x,t) = \phi(x/t)$ we have

$$\frac{\partial c}{\partial t} = -\phi'(\omega)\frac{x}{t^2}, \quad \frac{\partial c}{\partial x} = \frac{\phi'(\omega)}{t},$$

$$\frac{\partial^2 c}{\partial t^2} = \frac{\phi''(\omega)\omega^2 + 2\phi'(\omega)\omega}{t^2}, \quad \frac{\partial^2 c}{\partial x^2} = \frac{\phi''(\omega)}{t^2},$$

and the given ordinary differential equation readily follows.

14. If $f(c) = f_0$ then we have

$$\frac{\phi''(\omega)}{\phi'(\omega)} = \frac{2\omega}{(f_0^2 - \omega^2)},$$

so that

$$\log \phi'(\omega) = -\log(f_0^2 - \omega^2) + \log C_1,$$

and the required result can be deduced. From

$$\phi'(\omega) = \frac{C_1}{2f_0}\left\{\frac{1}{(f_0 + \omega)} + \frac{1}{(f_0 - \omega)}\right\},$$

we obtain on integration

$$\phi(\omega) = \frac{C_1}{2f_0}\log\left|\frac{f_0 + \omega}{f_0 - \omega}\right| + C_2.$$

15. If $f(c) = c$ we have

$$(\omega^2 - \phi^2)\phi'' + 2\omega\phi' = 0,$$

which is readily seen to be invariant under the given group. Thus from $\phi = \omega\psi$ and

$$\phi' = \omega\psi' + \psi, \quad \phi'' = \omega\psi'' + 2\psi',$$

we obtain

$$(1 - \psi^2)(\omega^2\psi'' + 2\omega\psi') + 2(\omega\psi' + \psi) = 0,$$

so that

$$(1 - \psi^2)\left(\frac{d^2\psi}{d\tau^2} + \frac{d\psi}{d\tau}\right) + 2\left(\frac{d\psi}{d\tau} + \psi\right) = 0$$

and the given equations follow on using $p = d\psi/d\tau$.

16. (i) $\xi(x,t,u) = \alpha x t + \beta x + \gamma t + \delta$,

$\eta(x,t,u) = \alpha t^2 + 2\beta t + \kappa$,

$\zeta(x,t,u) = -(\alpha t + \beta)u + (\alpha x + \gamma)$,

where $\alpha, \beta, \gamma, \delta$ and κ denote arbitrary constants.

(ii) $\xi(x,t,u) = \alpha x + \beta t + \gamma$,

$\eta(x,t,u) = 3\alpha t + \delta$,

$\zeta(x,t,u) = -2\alpha u + \beta$,

where α, β, γ and δ denote arbitrary constants.

23. (ii) From the relations $x = x_1 - \epsilon t_1$, $t = t_1$ and

$$\frac{\partial u_1}{\partial x_1} = \frac{\partial u_1}{\partial x}\frac{\partial x}{\partial x_1} + \frac{\partial u_1}{\partial t}\frac{\partial t}{\partial x_1}, \quad \frac{\partial u_1}{\partial t_1} = \frac{\partial u_1}{\partial x}\frac{\partial x}{\partial t_1} + \frac{\partial u_1}{\partial t}\frac{\partial t}{\partial t_1},$$

we may deduce

$$\frac{\partial u_1}{\partial x_1} = \frac{\partial u}{\partial x}, \quad \frac{\partial u_1}{\partial t_1} = \frac{\partial u}{\partial t} - \epsilon\frac{\partial u}{\partial x}, \quad \frac{\partial^3 u_1}{\partial x_1^3} = \frac{\partial^3 u}{\partial x^3},$$

so that

$$\frac{\partial^3 u_1}{\partial x_1^3} - \frac{\partial u_1}{\partial t_1} - u_1\frac{\partial u_1}{\partial x_1}$$

$$= \frac{\partial^3 u}{\partial x^3} - \left(\frac{\partial u}{\partial t} - \epsilon\frac{\partial u}{\partial x}\right) - (u+\epsilon)\frac{\partial u}{\partial x}$$

$$= \frac{\partial^3 u}{\partial x^3} - \frac{\partial u}{\partial t} - u\frac{\partial u}{\partial x},$$

and therefore the equation remains invariant under the given group. The functional form is obtained by solving

$$t\frac{\partial u}{\partial x} = 1,$$

which gives $u(x,t) = x/t + \psi(t)$ where ψ denotes an arbitrary function of t. On substitution into the given equation, we obtain

$$\psi'(t) + \psi(t)/t = 0,$$

and therefore $t\psi(t) = $ constant and the given solution follows immediately.

SUMMARY OF RESEARCH AREAS

Although the research areas listed below, have been discussed to a varying extent throughout the book, this summary is intended to focus possible research avenues, where additional work would be useful, for the student interested in pursuing the subject further.

Ordinary differential equations

(i) Abel equations of the second kind (pages 7-12 and discussion below).

(ii) Lie's fundamental problem (pages 51, 52, 63-66).

(iii) Differential-difference equations (pages 4, 74-77).

Partial differential equations

(iv) Non-linear diffusion with $D(c) = c^{-4/3}$ (pages 122, 138, 155).

(v) Classical groups for important equations (discussion below).

(vi) Application of group approach to moving boundary problems (pages 104-107).

(vii) Non-classical groups for the diffusion equation and others (pages 116-118).

(i) In the analysis of both ordinary and partial differential equations we have seen that the group approach tends to lock into Abel equations of the second kind, for which the standard form is

$$y\frac{dy}{dx} + a(x) + b(x)y = 0,$$

and for which further solution techniques, for particular classes of functions $a(x)$ and $b(x)$, would be highly desirable. We must bear in mind that the transformations recommended by group invariance, although reducing the order of the equation, may well induce other complexities. The examples at the end of Chapter 5 provide good illustrations of this. Take the equation of Example 5.10 with α zero, namely

$$y'' = \beta e^y.$$

If we effect one integration by means of the substitution $z = y'$ and a second from the substitution $\omega = \exp(-y/2)$ then we can show that the general solution is

$$y(x) = -2\log\left(\sqrt{\frac{2\beta}{C_1}}\sinh\frac{(\sqrt{C_1}x + C_2)}{2}\right),$$

where C_1 and C_2 denote arbitrary constants and we are assuming C_1 is positive. However if for $\alpha = 0$ we follow the strategy of Example 5.10 then from (5.45) we obtain the following Abel equation of the second kind

$$up\frac{dp}{du} = p^2 + up - 2u^2 + \beta u^3,$$

which at face value is not of a standard soluble type and yet from the above general solution we must be able to solve this equation. Thus Abel equations may not be as fearsome as we presently think.

(ii) We know that the first order ordinary differential equation $y' = F(x,y)$ can be invariant under an infinite number of one-parameter groups. If we differentiate this equation with respect to x, we obtain either of the second order equations,

$$y'' = \frac{\partial F}{\partial x} + \frac{\partial F}{\partial y}y', \quad y'' = \frac{\partial F}{\partial x} + F\frac{\partial F}{\partial y},$$

which are invariant under at most eight one-parameter groups and which moreover can be determined systematically. The question arises as to whether groups for one or both of these second order equations can be utilized to integrate the original first order equation $y' = F(x,y)$.

(iii) In a recent research article (volume 38 of the IMA Journal of Applied Mathematics (1987), pages 129-134) Shigeru Maeda applies the similarity method to ordinary difference equations. As noted in Chapter 1, it would be desirable to extend the group method to differential-difference equations.

(iv) In Chapter 7 we showed that the non-linear diffusion equation (7.3) with diffusivity $D(c) = (c+\beta)^{-4/3}$ admits a wider class of group invariance. Moreover Problem 10 of Chapter 6 shows that the index $4/3$ is also critical for the diffusion equation with an inhomogeneous diffusivity. Combining these results we might expect that the equation

$$\frac{\partial c}{\partial t} = \frac{\partial}{\partial x}\left\{\left(\frac{\alpha x + \beta}{\gamma c + \delta}\right)^{4/3}\frac{\partial c}{\partial x}\right\},$$

plays a priviledged role in non-linear diffusion theory. While a good deal of research has been undertaken on the corresponding equation but with index 2 (namely equation (7.53)), very little work has been done on the above equation.

(v) For partial differential equations in general there is still a good deal to be done and it is not difficult to find interesting examples which are worthy of analysis. In many cases the classical groups leaving the equation invariant are known, but the resulting functional forms and ordinary differential equations have yet to be studied in detail (see for example Problems 20 and 21 of Chapter 6 and Problem 16 of Chapter 7).

Summary of Research Areas   201

Throughout this book we have only discussed equations with a single dependent variable but there are many equations in Applied Mathematics with more dependent variables. For example, coupled reaction-diffusion equations such as,

$$\frac{\partial c_1}{\partial t} = D_1 \frac{\partial^2 c_1}{\partial x^2} - k_1 c_1 + k_2 c_2, \quad \frac{\partial c_2}{\partial t} = D_2 \frac{\partial^2 c_2}{\partial x^2} + k_1 c_1 - k_2 c_2,$$

where D_1, D_2, k_1 and k_2 all denote positive constants. It would certainly seem worthwhile examining implications of classical invariance for such systems.

(vi) While we have briefly mentioned the relevance of the group approach to moving boundary problems, we have by no means exhausted the subject and the area emerges as one of the most potentially rich and promising areas for study by group methods and moreover it may well be the only approach likely to lead to analytical solutions to such problems. In particular, moving boundary problems are frequently characterized by conditions at the moving boundary $x = X(t)$, such as

$$c(X(t), t) = 0, \quad \frac{\partial c}{\partial x}(X(t), t) = -\dot{X}(t),$$

where $c(x, t)$ denotes the concentration or temperature and we would like to find general classes of groups leaving such equations invariant. For futher details and references we refer the reader to Hill (1987).

(vii) Very little work has been done on non-classical invariance of partial differential equations. For example non-classical groups for the diffusion equation are governed by the system (6.80), namely

$$\frac{\partial A}{\partial t} = \frac{\partial^2 A}{\partial x^2} - 2A \frac{\partial B}{\partial x}, \quad \frac{\partial B}{\partial t} = \frac{\partial^2 B}{\partial x^2} - 2B \frac{\partial B}{\partial x} - 2 \frac{\partial A}{\partial x},$$

which would certainly seem to warrant closer attention. Moreover there are many other important equations such as the non-linear Burger's equation and the Korteweg-de Vries equation

$$\frac{\partial u}{\partial t} + u \frac{\partial u}{\partial x} = \frac{\partial^2 u}{\partial x^2}, \quad \frac{\partial u}{\partial t} + u \frac{\partial u}{\partial x} = \frac{\partial^3 u}{\partial x^3},$$

for which particular non-classical groups and their resulting solutions might correspond to important and new physical phenomena.